Quasicrystals

and quasi drivers

Antony J. Bourdillon

UHRL
P.O. Box 700001, San Jose, CA 95170

AuthorHouse™
1663 Liberty Drive, Suite 200
Bloomington, IN 47403
www.authorhouse.com
Phone: 1-800-839-8640

First published by AuthorHouse 2/24/2009

ISBN: 978-1-4389-5589-6 (sc)

And by UHRL:
ISBN: 978-0-9789-8391-8

Printed in the United States of America
Bloomington, Indiana

This book is printed on acid-free paper.

Contents

Part II

---

[*] Endnote numbering starts in part II, replacing footnote numbering in part I

1. Introduction

There are two types of driver. One is the chemical driver for an atomic structure[i], the other is a set of puzzle builders. The latter figure in part I of this book. Details are given in footnotes, but the first part is more general than the example in part II.

Quasi drivers? Beyond revealing the structure of 'quasicrystals', this book extends to quasi reviews – of various kinds: referee reports; their effect on conduct in various branches of physics; and on quasi science in general. What physics thinks to be its virus is actually its prophylactic. Smolin[ii] touched on the problem when describing his experience in the theory of fundamental particles. Other disciplines, including quasicrystals and X-ray lithography, are in similar 'trouble'. When physics or science lets us down, we file to the internet. This is an uncensored relief from a limited and all-too-human system. Times change and customs vary. Occasionally now, reviewing is open, but physics has retained its ancient aura. Part I describes some shortcomings, making necessary its sequel.

Part II is a recent solution for the otherwise problematic structure of quasicrystals. For 25 years, this has been the most fundamental unsolved[iii] , [iv] structural problem in condensed matter physics. It has attracted researchers from

---

[i] Appendix C, page 99.
[ii] Lee Smolin, *The Trouble with physics – The rise of string theory, the fall of a science, and what comes next,* 2007, Houghton Mifflin, ISBN 13-978-0-618-91868-3
[iii] Senechal, M., What is a Quasicrystal? *Notices to the American Mathematical Society,* **53,** 886-887 (2006).
[iv] Steinhardt, P.J. and Jeong, H.-C., *Nature,* **382,** 433-435 (1996)

many branches of science; but it is also beset by long-standing faults that raise broader questions.

What is science? It is like a circus because it has performers and institutions, but different because it has laboratories, conferences and journals. It is like news media because it has accepted truths or 'facts', but different because they are tested by laboratory experiments. It is like politics because it has secrets and lies and power, but different because experiments can be checked, and its power is derivative on those politics that finance its laboratories. It is like religion because it has hierarchies and priests, but different because, in the long run, experiments rule. Those 'priests' are the qualified or recognized scientists that do the experiments, develop their theories, administer their journals and say what science is. This book is for the long run.

What you're reading is like a book because it has pages, print and a binding, but different because it has serious content. In desktop days, it is an example of its own doctrine. The book is written by an author and frequent referee to many journals, but different because it is uncensored. As the story unfolds, you should see there is trouble in science, as with the 'physics' described by Smolin - where theory is divorced from experiment by a science networked in self-interest.

The purpose of the following chapter is not to describe comprehensively the structure of journals; but to give enough background to support the criticism and example of censorship. Galileo lives still – now, and on the internet. What drives this show – true chemical forces or a closed system of nameless referees?

2. Peer review

". . if there be any difference of opinion, and it seem good to let it appear, let the reasons be put forward with modesty and charity, with intent to establish the truth, and not that they may appear to have the upper hand [v]. "

Underscoring all interpretations, journals accept papers for publication by the method of 'peer review:' Peer:

1. An equal in civil standing or rank...
2. One who takes equal rank with another in point of natural gifts or other qualifications...
3. One who is associated or matched with another...
4. A member of one of the degrees of nobility in the United Kingdom...
5. In generalized sense: a man of high rank in any country, state or organization...[vi]

Lord or beggar? This range of meaning allows interpretations wider than the Pacific ocean. You can imagine fellows in an Oxbridge common room telling each other how equal they are, and how each just invented five reasons for bouncing each of five papers. Imagine them, 'peer review is a way for scientific papers to be accepted by equals in the scientific community.' 'Of course we only review, we don't censor. Our own interests never conflict. We do it only as duty.' How do journals work?

Some journals are owned for profit and others by independent learned societies, or charities, non-profits or

---

[v] The Rules of the Society of Jesus (Roehampton: Ex Typ.[ographia] Manresana, 1894, pp. 29–30)..
[vi] Oxford English Dictionary, 1971, reprinted with permission.

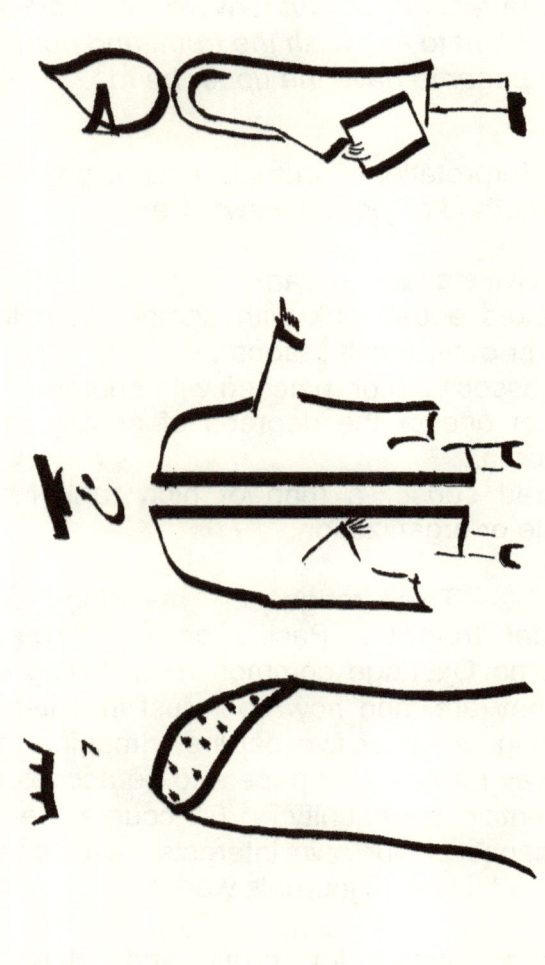

The review: "You're a *peer*. I'm a *peer*. He's a *pear*."

whatever. Well known scientific publishers are the American Physical Society and the Institute of Physics, London. The journals are supported partly by sales, to libraries especially, including university libraries, but also to individual investigators by subscription. Each journal has an editor and most have an editorial board that makes policy and helps in ameliorating disagreements between referees and authors. Naturally, editorial boards generally support editors. The journals need authors and readers, so both editors and boards have to keep an eye for rankings. Simply put, the rankings depend on how often a journal's articles are cited in other journals.

Authors (usually) need journals so there is a complementary relationship. Authors can select between journals before submitting. Typically, each journal will publish its scope and publishing policy, both on printed issues and through electronic media. Authors are expected to match their papers to journal scope. Typically, an author, working in a chosen field, will build a developing relationship with a compatible journal.

The best journals publish policies for resolving disagreements. They don't have to keep to the policies, but they are there to encourage potential authors.

An editor will keep a list of colleagues (buddies) that will give him or her advice on whether to publish particular papers. These are the semi-notorious referees that all we scientists have disagreements with from time to time. Editors often invite their authors to join their pool of referees. They then serve, in turn, to spread the load of peer review. The editor has the right of final decision.

If an author or referee has an unusual interest in a paper, it needs declaration. Authors from drug companies are particularly vulnerable. An author is often given the opportunity to recommend referees and deny others. This book is written chiefly from the perspective of an author. When he was invited to join an editorial board Author weakly accepted but was saved by a change of management. He has since avoided them, chiefly for lack of commitment as shall be explained.

How do authors submit their papers to scientific journals? Within working memory, it was normal for an author to submit three typed copies of his paper to an editor: one for the editor and two for the referees. If the referees both approved the paper, it could be printed. Usually a change, proposed as an improvement, would be recommended or required. Referees generally appear smarter if they suggest something. Their suggestions are more often trivial than improving but it's best for an author to play along. Nowadays, submission is invariably electronic. Copies are easily made and quickly circulated. The refereeing remains otherwise archaic.

Unfortunately, peer review is - more often than not - both biased and competitive. You'll find examples in the following chapters.

3. When it works

There are two types: stamp collecting, and physics. Peer review works well for collectors and you can see why. Imagine you are the referee: 'Here's another stamp. I don't think I've seen it before. No axe to grind. No reason why not. Don't make trouble. Editors need papers. So do authors. Libraries need journals. Publishing pays. Easiest to put it in, physics aside. Nobody will object. OK.'

But what if the paper overturns paradigms; steps on toes; 'conflicts with my recently published proceedings'; 'offends my friends?' No need to be specific: 'the readership won't like it'; 'It's been done before,' (no need for me to say where or when); 'there's a conflict with [irrelevant] facts'; everything except 'My paper is not cited and therefore I don't, subjectively, want this one'. Next chapter for the bad news; first the good news.

The majority of scientific papers form an erudite collection with little to say for or against anything much. As a well-known author and prize winner, Richard Feynman gave this general advice, that when your paper is rejected, don't dispute it - send it to another journal[vii]. It's a point of view, valid perhaps for some authors in some cultures and in some circumstances. However, others will sometimes do better without this advice. Following, are some examples. The first objection to the advice is delay. This is implicit in submitting for the second time, and this often requires extra work in re-editing of the manuscript. The delay is doubled when it is repeated. It is compounded when novel papers

---

[vii] Feynman, Richard P., Leighton, Ralph, and Hutchings, Edward, (1986) *Surely you're joking, Mr. Feynman*, Mass market paperbacks.

are rejected multiple times as referees unite in excluding upsetting ideas, however valid their contributions. What options then? Sometimes there are advantages in disputing referees for honest and noble reasons, independent of the fruitless rectitude in redressing an insult.

The disputed cases that follow are not all strictly verifiable; but that doesn't matter, since every scientist that has been born without a silver spoon in his mouth will recognize similar events. Proving the detail is therefore not of the essence. Besides, all good fiction is biographical.

Case 1. The adjudicator who did not check his facts and was overruled by the board:

Author described a painstaking study made on sophisticated instrumentation. The study described a new way of measuring atomic structures about selected species of atoms, quasi independently of all the other atomic species present in the solid. First reviewer wrote that it was a good paper and should be published: 'thoughtful and significant.' Second reviewer said the work had been done before and cited an obscure laboratory report. Editor appointed an 'adjudicator' to decide between the two reviews. Adjudicator sided with rejection but made a series of unforgivable howlers.

When the reviews reached Author he called Editor and asked if a response would be taken seriously. Editor said it would go before the editorial board at their next meeting. After great effort needed to track the laboratory report, author replied that:

- Adjudicator had evidently been influenced by the second review but that, from internal evidence, he had proved incapable of checking facts (5 obvious inconsistencies in the adjudication were cited – of the type, 'Author used the word x on page 5 where he should have used y'; where in fact Author used the word x neither on page 5 nor anywhere in the paper!).

- Author pointed out, in particular, that the previously cited work had attempted to detect the signal, currently described, but 'could obtain no blackening of the photographic plate.' This was 'a long way from proving the feasibility of site selective EXAFS,' protested Author!

- He wrote, 'it is not my intention to referee the adjudicator, but to show that further examination is required.'

As a triumph for rationality, the advice of the first reviewer was accepted by the board, and the paper was duly published. Author was even better at letters than papers.

How did this happen the way it did? As usual, the identities of adjudicator and both reviewers were withheld from Author, but this does not affect the obvious conclusions to be drawn. Unchecked arguments were used by two referees to attempt to bounce a painstaking, consistent and competent piece of research. Reviewers generally have the power to exercise unsupportable generalities to reject publication. Personal vendetta, why not? In this case, it may not have done them any good in the long run. Editors have power to reject generalities but rarely do so.

Friends are more reliable than enemies, but sometimes rivalry born of intuitional disagreement on scientific direction is inevitable. Reviewing should be rational and therefore open. Author is at a disadvantage when he cannot evaluate a referee who conversely knows the author's experience, facilities and limitations. Rationality is a duty for editors.

From the viewpoint of an author, journal review allows interesting comparisons with Patent office review. Most patents are written by professional agents. Then the author is a kind of technical middle man. But in the United States, Congress has encouraged individuals to disclose inventions so that they can go into the public domain without prejudicing the economic return that the inventor should expect. An inventor's advantage in filing individually is that fees are halved. Moreover the Patent office is obliged to give extra help to individual applicants.

Between journals and patents, there are differences in convention. Do-it-yourself advice[viii] is that it is fruitless to dispute the Patent office but, as simply as possible, do what they say. Sometimes you can get your way in spite of the negative advice. First file a Patent Application. Patent Office will dispute the claims on the most general grounds it can think of. Don't be disheartened, they always do this. In responding to journals you make compromises with the criticisms by reviewers. Don't do this with the Patent Office or they will say you are disqualifying the Application by filing new material. In patenting, the filing date for the material – especially for the figures and claims – is significant. Assuming the invention is good and that you have done your background searches properly, just make minimal changes. According to the rules, you are allowed to change the patent claims in the first round but not afterwards. At each stage the application goes first to a Legal Instruments Clerk who passes it, as she sees fit, to the Examiner. This is the guy who says Yeh or nay. You are allowed to telephone him at critical times. The hierarchy is more complicated than shown. It is described on the web page of the United States Patent and Trademark Office, but this is a cut-down version that Applicant sees:

[viii] *Patent it yourself*, Nolo, www.nolo.com

Commissioner of Patents.
  Examiner
    Examiner's supervisor
      Supervisor of legal instruments clerks
       Legal Instruments Clerk

Case 2. Mistaken Legal Instruments Clerk.

What happens if Legal Instruments Clerk appropriates the role of Examiner and refuses to pass on your application?

In this case Author made too many changes in the first round. Clerk judged that new material had been introduced but that the application would have passed without some of the changes. On the second round, author withdrew all of the changes reverting to the original script excepting the one claim withdrawn on the first round. Author showed that each change made in the first round was not new material but a repetition of material already contained in the Application. Therefore, by returning to the original - which examiner had described as acceptable except for the 'new material' - Applicant claimed the Application should be allowed. Not so for Legal Instruments Clerk, who said that the claims had been changed. Now, since one had changed in the first round but none in the second, the clerk was in error. Applicant pointed this out to Examiner who said, 'I see what you mean', though, by appearance, he could do nothing else. Applicant was obliged to pay a significant fee to continue the examination. Applicant filed again a correction pointing out the error. Legal Instruments Clerk objected with more errors and put the application on the shelf. What she claimed to be errors were recent rule changes that were implemented so close to the publication date as to be impossible for Applicant to follow. The changes also contravened specific instructions previously given in writing by Clerk. Applicant called Legal Instruments Clerk pointing out the errors and unpublished rule changes. She was unmoved. When asked whom to complain to, she said, "The Commissioner of Patents." After 6 months, Applicant spoke to Supervisor of the section dealing with the type of patent. He said he could do nothing unless Supervisor of the legal instruments clerks approved, which she didn't. Applicant wrote to the Commissioner of patents, copied to the Examiner, saying among detailed evidence

- There is no mechanism for correcting errors made by Clerks.
- Supervisors do not have the authority to see due process.
- The goal posts keep moving.

Within 48 hours of mailing, Applicant received a phone call from Supervisor saying, 'We have decided to waive all rules!' Now Congress makes laws, and the role of Patent Office is to make derivative rules and implement them. Imagine the Commissioner of Taxes calling to say, 'We've decided to waive all your taxes.' Equivalent? To get his patent, Applicant had only to respond once more, after speaking again to Examiner.

Patent Office is required to give special help to individual Applicants. This time Applicant wrote the 'request for conditional assistance' clause at the beginning, instead of the end, of his fifth revision. Within two weeks of the phone call, but three years after the filing, issuance was allowed.

Interacting with the patent office is instructive because of differences from journals. Firstly, Applicant can get to speak to all the players on a limited basis, keeping interactions occasional and brief. From these conversations, an applicant can get the measure of the players – whether reasonable, defensive, obstructive etc. – even though he doesn't normally get to meet them. With journals, speaking is rare. On the other hand, whereas correspondence with the Patent Office is formal, expensive, slow and therefore uncommon; emails to a journal administrator are typically informal and routine. The measure is harder: at best Author can guess who a referee is; but Author is usually in the dark. Referee, by contrast can often remember hearing Authors speak at conference and may know them or remember meeting them, even have confronted them. At the same time, an editor may be known to both referee and author and so can play mediating, as well as decisive, roles.

Suppose an author has a breakthrough discovery with a paradigm shift in an area of research with massive commercial implications. How do journals react?

Case 3. Solid, but not impregnable, opposition.

Author wrote a paper, ground breaking. He was comparatively new to the field, but the research was essential to his new laboratory. Author filed a provisional patent before submitting the paper to a journal in Japan. Referees wrote that the work was not new. Years later it turned out that another patent was filed in Japan on the exact topic. This should be valueless, even if disputable, owing to priority of the first patent. It was so difficult for Author to prove that unknown referees misused the submitted paper that it was not worth trying. Author had no recourse to professional redress, where submitted papers should be held confidential until published. Other examples will be given of misuse of confidential information. In this case apparently, rejecting publication aided Referee in improper filing for intellectual property. Contrary to the report, the work was of course new, as the patents proved – as evidently as the impropriety was then unprovable. Even later, prior anonymity of reviewers carried virtual indemnity.

What did Author do? He followed Feynman's advice. Choose a journal you know, having a different set of reviewers. Let it be not cutting edge in the field, so that competition will be less probable. Author published a series of papers without further problem. Further patents followed. With each paper the concepts became more established and harder to reject. As for redress, that is commerce for future dispute in courts of law, whatever the science.

It is more common than not that referees are in competition with authors, which is one reason why they should be open and signed. That way, referees' occasional contributions could be acknowledged.

Case 4. Author proves a point.

Author mailed a paper to a journal, knowing the editor in country A. Editor transferred paper to Sub-editor in country B. Author assumed paper would bounce for personality reasons. Instead of withdrawing, he immediately submitted to a second journal of excellent reputation. This is against protocol but Author judged this a special case. Referee reports returned simultaneously: the second accepted without change, the first rejected – as predicted. Sub-editor did not hide his annoyance. His expression was at the breach of protocol; though obviously the annoyance was at the successful insult.

Who's the peer, who the pear?

4. When it doesn't

*Peer review is a discipline for young
scientists, but censorship for the creative.*

When refereeing doesn't work is when it is most needed – to help readers understand the general acceptability of new material.   Referee approval doesn't matter, of course, to genuine scientists, who prefer to make up their own minds, without restraints of censorship or approval.

In the age of the internet, nothing is too novel to publish, provided you're happy with going over the top of censorship. But when you feel restricted by scientific conventions, taken out of the ark of history, you have some new options.   The following case refers specifically to part II of this book.

Case 5.   The Galileian syndrome – too novel to publish.

In this case Author was right; Referee agreed he was right; reader can *see* he is right; but Editor would not publish. Like the Pythagorean star, the history has five parts.

None of the real issues were addressed: are the discoveries true; are they significant; are they professionally written?

Author had first submitted a novel paper to a journal which had already printed a series of his papers one of which was used in promotions. The journal had published many papers on the topic of quasicrystals[ix]. These constitute a new type of solid, in between a glass and a crystal, but in fact neither of them. They are a puzzle to mathematicians because of the way their atoms fill space with five-fold symmetries that are not allowed in crystals. Here, as is explained in part II,

---

ix http://en.wikipedia.org/wiki/Quasicrystal

15

we had mathematicians wagging the dog with clear experimental evidence being denied publication for (political) convenience. First journal had held the paper for five months, while Editor failed to find a referee willing to advise. Without benefit of doubt, he declared it out of scope. That was part 1.

A second journal, known for multidisciplinary interests, rejected the paper writing:

> 'Thank you for your letter asking us to reconsider our decision on your manuscript......
> 'We appreciate the points you raise, and are not questioning the validity of your work or its interest to others in the field. Nevertheless, our view remains that your manuscript is of insufficient immediate interest to our broader readership in terms of firm new and general physical insights to justify its publication in XXX...'

Polite but, true or false, general relativity could have been rejected for the same reason – and hardly true in the light of a large number of papers already published on the same topic.

Given the unfortunate delays, author responded by writing the basic ideas onto a personal website, now copied to part II of this book. The website was referenced from various other public websites. That was the third part. Then he wrote a subsidiary paper for yet another journal: part 4.

This third submission was prefaced by a personal approach to a sub-editor. He went silent when he first understood that the original pattern was not icosahedral[x], even though it had been accepted as such by 'hundreds, if not thousands'[xi] of researchers for 25 years! Author made clear why the

---

[x] appendix D, p. 102

[xi] in the words of the single referee who knew he was wrong *in fact,* admittedly, and unashamedly!

diffraction is not Bragg diffraction as had been widely assumed[xii]. What difference could it make to a rolling freight train that every word in the paper and every concept was novel? For the first time, eleven months after first submission, an editor was able to find a referee willing to give advice. In a mood of denial, his rejection was expressed in factual error mixed with banal generality, but the reference had become complicated by part 5 which was a further subsidiary submission to a related journal: 'Ouch, my toes! Not this', you could have heard Referee startle.

'I strongly recommend this manuscript for rejection by YYY. The HRTEM [high resolution transmission electron microscope] work upon which the paper is based is from the 1980s. Quasicrystal research has moved on since the 1980s in terms of structural models, as the recent set of papers from ICQ9 published in ZZZ will show. I commend these to the author, as well as recent papers on quasicrystals published in the open literature. At the very least, the author should have compared his model with the current models of quasicrystals in a paper worthy of publication in the open literature. In any case, there is no simulation of HRTEM images to corroborate the 'look-see' interpretation of the HRTEM images – simulations of HRTEM images have also moved on since the 1980s. Furthermore, HRTEM is only one of a number of techniques which have been used to determine the atomic arrangement within quasicrystalline materials. The author should take particular note of recent published work based upon X-ray diffraction and neutron diffraction data....'

Confused bombast of course, but there is no arguing with opinionated reviews. Actually:

- The data *is* from 1987, but scientific data does not age, otherwise palaeontology would not be a science. The data is valid data.
- Referee displays logical confusion: whatever their dates, models and data are independent. A new model can be justified by old data.

[xii] appendix B, p.95

- Besides, sometimes old data is not interpreted for a long time as in the CBED[xiii] in this case[xiv].
- At the very least, Referee could have noticed that this model differs by being driven. Referee's commendation for knowledge that is readily available on the internet, wastes words.
- *As appropriate 'pear review', Referee is commended to read the papers he agrees to review!*
- What use in repeating what hadn't been read? Author parted company with other models at the starting post: Author's model defined a single driving force[xv]; 'current models' jumbled jig-saw pieces like monkey puzzles. This difference was clear, but Referee refused to notice it.
- It is unprofessional to make unnecessary controversy – 'current models' were correctly mentioned in passing. Could Referee imagine Author copied?
- The HRTEM image (figure 7 below) was originally published without simulations:
  - Simulations manipulate and falsify
  - It had been explained that this particular data was acquired under special conditions suitable for direct interpretation
  - The original data had been published in a more highly ranked journal without simulation, at a time when simulation techniques were already well developed
  - Not only have developments in simulation been minor; but special and untested adaptations would be needed to account for quasicrystal diffraction in place of conventional Bragg diffraction used for crystals. Multislice methods[xvi] that use normal dynamical diffraction do not apply!

[xiii] Convergent beam electron diffraction.
[xiv] Bourdillon, A.J., Phil. Mag. Lett. **55** 21-26 (1987)
[xv] Appendix C, p.99
[xvi] http://www.hremresearch.com/Eng/simulation.html

- His unusual choice of acronym implies Referee is not even a transmission electron microscopist (expletive deleted)!
- Referee chose to assume Author was not aware of X-ray diffraction. Never mind that:
  - X-ray data had been reported years before.
  - The first reports of X-ray diffraction were uncharacteristically[xvii] wrong.
  - The data mentioned was of dubious significance, as Author was aware.
  - Significantly, Referee omitted to say what he imagined was important about it.
- What *will* be published is an old subterfuge for the gullible – why total denial?
- The atomic arrangement had not been 'determined': several review articles, some cited, have it written that the structure is not known.
- The law of universal mendacity[xviii] teaches that what was said is false by reason of being said. When no reason is given, the meaning resides solely in what Referee opposes.

How can one be sure that Referee had no part in the work he wanted cited. Anonymous refereeing makes tracking impossible. Such generality in the face of extensive and consistent novelty is counter-productive. Hence the email (next page) to Editor.

Without knowing about the experience and interests of Referee, how to assess its accuracy? If not you, can Editor for whom the review was written? Why should he even forward it to Author, when its provenance is so clouded?

---

xvii Pauling, L., *Letters to Nature,* **317,** 512-514 (1985). Unfortunately the author chose not to evaluate electron diffraction data. A-Pauling (3 new meanings)?

xviii Yang, B.D., *The law of Universal Mendacity,* AuthorHouse, 2004

What happened to rational argument and openness in science?

Gentle reader, is there an untutored eye that cannot match the patterns that are common in figures 6, 7a and 7b below[xix]? Let him or her know the scales also match.

Before inspiration came this deep sigh:

> 'Dear [Editor],
>
> 'You realize, don't you, that [the assistant editor] and your referee are two of the 'hundreds if not thousands of researchers' who all got it wrong IN FACT, admittedly, and now unashamedly! If you too consider it 'trivial' when the pattern is not icosahedral, there are questions about your mental health.
>
> 'What a hoot! More details on www.UHRL.net …..
>
> 'The work is published [on the internet], so there is little to be done, - but there are wider issues…....'

…as described in this book. This is the kind of letter you might write to an editor if you think he might have the personality needed to sort it out. As you will infer, the referees are in denial and in disgrace – proved wrong by experimental fact, but trying, as they will, by unsubstantiated rejection to protect themselves.

By this stage Author realized he needed journals about as much as any other serious scientist – and not at all incompetent journals of whatever reputation. It has become so easy to epublish and to search, when open source is so cost effective and accessible, that journals have become irrelevant. Author collated all four papers and posted a new

---

[xix] On pages 63-67.

webpage [xx] in communicative format with the following advantages:

- Format *is* important. Scientific journals are staidly stylized with the result that information is hidden. By contrast, web graphics get straight to the point.
- This is one reason why readers, including referees, often do not get the message.
- Updates are readily made,
- with no delays greater than 10 minutes, and
- no censorship.

Case 6. Sustainable obstinacy.

This is a case of how editors spin themselves into webs trying to avoid decisions.

Author had made a fundamental development of a technique of great commercial importance. That it was commercial is unimportant, since commerce insulates itself from science, which is rarely used when not of benefit to market leaders, startup companies being exceptional. To get published in a journal, Author had to overcome a virtual cartel of referees. On the way he proposed a review of the field and was commissioned by Editor. Meanwhile Author prepared a research paper describing latest results and submitted it to the same journal. There followed a series of referee reports and rebuttals requiring Editor to personally evaluate the paper. This is what he decided. Author should write the review and incorporate the paper into it. But when Author asked about the review process, he was told no change. The forthcoming review is still on the published page of future features, and that started five years ago!

[xx] http://www.UHRL.net, copied to part II of this book.

Case 7.   Unprovable theft.

Scientists will always complain of theft of ideas. Sometimes coincidences are so remarkable that they cause more than suspicion. The ease with which theft occurs makes the journal system corrupt.

Author had written a paper describing a new phenomenon in a widely researched topic. Within one month of submission, the same phenomenon was described with fanfare in the press by another scientist. The two groups competed in different countries for different funds, but that is not important. The same discovery, made in two places at the same time can happen. What causes suspicion is what happened next. In spite of repeated reminders, Editor responded after a delay of one whole year:

'Your paper is accepted without change.'

No explanation was given for the delay. Now it is rare for a paper to be accepted without change. It is rarer for the refereeing process to take more than three months. If there had been theft, the delay was an excellent cover. After another year the paper was printed.

Is everything printed by journals out of date?   Did Author have any means for redress? Bringing an action against an unknown referee was unlikely to be effective. Editor could be expected to protect himself by obstruction and cover-up. Author decided to move on, as he had a lot more discoveries still to make. With responsible and open refereeing, if it ever comes, Author's chances would be better.

5. Electronic publishing.

*When you stop stamp collecting, you shouldn't
expect journals to support you.*

The internet.

It is no longer *necessary* for a scientist to publish in journals.
Desktop publishing is easy and economical - with ready
search facilities.   In fact the best work is posted on the
internet independently of the journals, and these are mostly
electronic anyway.   Epublication is virtually instantaneous.
By comparison journals are lumbering; conferences are out
of date before they meet, and their proceedings history.

The personal web page has unrestricted format and so can
be designed for better communication. It is updatable within
10 minutes. Why then submit to journals at all? For grants?
For grant givers? For jobs? For academic promotion? For
prizes? For circulation and advertising? For the stamps that
don't justify personal web pages? For old times sake? But
when your interest is science, why hassle?

A smaller difference between journal articles and personal
publication is that the first is restricted to people with access
– typically through university library subscription, but also
through business, public library, or even personal
subscriptions. There are other differences too.

For those with long memories, good libraries carry journal
articles over 100 years old. Many libraries carry no science
printed in the current millennium. But they are repositories
for information. Journal articles are stored in university and
public libraries with care and accessibility that is not supplied
by other means. Once deposited, the articles are stored as

permanent records and not changed. This is a feature that many would regret to lose. But if your interest is in circulating your new discoveries, you can propagate them without bending to the limitations – in time and obstinacy – of refereed publication; without censorship, that is, by referees established in their own self-interest.

Wikipedia.com

How then does your work get noticed and cited? If you're a stamp collector, that hardly matters and the journals will most likely take it. If your paper has some novelty at an intermediate level, it could be problematic and may need individual consideration. It will quite likely pass the referees and can be published in journals in the usual way with the usual delays and disadvantages.   But if your paper is so outstanding that it is certainly not printable in journals, you can find ways of making it advertise itself. The internet has bibliographies and discussion groups that should be interested in it and these may be helpful. If you're ready for a little rough and tumble there is open source refereeing available in Wikipedia.   You post your discoveries in a cooperative way and argue back if part of your readership doesn't like it. Any one can edit the page. The discussion goes on record. The information is packed into categories: the article, discussion, history of entries etc. You are not supposed to post original research in an encyclopedia, but statements of fact and references to research are normal. Copyright material in not allowed on the open source site. The articles are informative, up-to-date, and as over-comprehensive as you would expect from multiple authors. They are also the most accessible of any scientific publications, and perhaps the most widely read. For novel research it is the best antidote to censorship.

How easy to make a web page

Original research can be posted onto a personal web page, with consummate ease and then kept under your own perfect control. Begin by finding a suitable hosting company. You can get free space, but if you need more control without advertising there are several that offer it for a modest fee, say $25 per annum. Doteasy makes it as you would expect from the name. You can write and edit a file to your liking using your favourite word processing software, provided only that it can save the file in html format. Then you need downloading software. Here again there are several options. A good one is cuteFTP from Globalscape, but you might save using open source. All you have to do is download your html file to your web page. Routinely it's easy, though if you have to get passwords right the first time round, use your head and persevere. The investment in time provides multiple rewards.

Book publishing

As you can overcome censorship on your scientific articles through the personal web page, you can do it in the wider scope in similar ways. Laptop publishing is easy for books too. There are firms that give you help with printing, circulation and advertising. You can get all this from AuthorHouse for a few hundred dollars before royalties come back to repay you. It is a small layout considering. Expensive print runs are avoided by print-on-demand technology. An alternative is your local printer. Write the file on your desktop giving a little thought to page format. After the print run, you'll have to use your own internet expertise to advertise and sell. Amazon Advantage is a resource. For a further smaller layout, you can assign your own ISBN numbers as a publisher and register your copyright. Otherwise, you can buy ISBN numbers. Victory is not to the weak.   No need to rely on this print run: for the latest information, Google it.

From this ease and desirable rapidity in methods of publication, it will be obvious that we live in the light of a new dawn.   With desktops, the gap between printing and journalism has closed.   That is why the history of publishing and its relation to journalism is topically interesting[xxi].   Many functions in editing remain the same, including quality control through checking facts, editing, design, agenda-setting, editorial style by selection of material and so on; but they are performed in different ways now.   What has not changed is the demand to make the published material compelling so as to instill trust.   That is why this book is necessary, together with the reforms described, before traditional editing loses its way.

[xxi] Fox, R, Eyewitness to history, 4 volumes, *Folio Society* 2008 seq.

6. Named and responsible

Or unnamed and irresponsible

Scientific publishing has lagged behind publishing in the humanities. As a typical example, it is thirty years since the *Times Literary Supplement* changed from anonymous to open reviewing. Book reviews are now routinely disputed and the disputes are often substantial and enhancing. Readers do not need to be told what to believe like controlled automatons. They make up their minds on the basis of evidence provided. By comparison, scientists are the automatons.

Book reviewing is not identical to refereeing of journal articles, but there are similarities. One is imperative: critics should be responsible for the views they give so that counter arguments can be evaluated on a level field and so that misdemeanors may be more accurately identified. Sometimes they may even be ridiculed, prevented or remedied. On the positive side, it is a matter of justice that the contributions of referees should be acknowledged. Nor is there strong evidence that honest referees would be less willing to review if the refereeing were open. There are professional benefits in early scientific reviewing in one's field of expertise.

If a referee refuses enlightenment, why test him? If a journal won't have your paper, go somewhere else. But if the method is not open and fair, discoveries will be hidden. We are back in the company of the great names of renaissance science. The discoveries can't be entirely hidden from the web, and it is now doubtful that they can be significantly hidden at all. Do what the best and the worst do: why not post your discovery onto the net.

27

Private posting may be a rational reaction for the individual, but is salvage possible in the public one? Few scientists actually have to read anonymous referee reports and they could refuse to do so. Many, not only referees, are involved in some way; but the chief responsibility lies with editors. Meanwhile, journals that fail to live up to their published policies should be down-ranked, but this could only happen with an appropriate supervisory body.

Failing that, and as it is human for journals to censor articles with which referees are allowed simply to disagree, authors will stop corresponding. Anarchy is the natural response to censorship. What kind of collector needs journals? Why spend time reading journals that have censored the best material? How long can print survive?

And if you are not yet convinced, compare the cost of this book with the cost of downloading journal articles. Word for word, they're a rip-off.

7. Quasicrystal or quasiglass?

Quasi icosahedral; quasi Bragg.

Contrary to universal[xxii] beliefs of 25 years, the original pattern is not icosahedral: the diffraction is not conventional Bragg; the model is uniquely simple; key is the driving force. Smolin[xxiii] has described studies in elementary particles, asking what would come after string theory. A further question is, 'What comes next in science?' In this book, a similar example has been made of quasicrystal research, so - to fill some gaps - here is an outline history. You can find book length descriptions elsewhere[xxiv]; this chapter is for general interest with specialized details in part II. The description supports the brief answer for the future of this field: nothing much – there is still work to be done on defects in the structure; but they can mostly be understood through its driving force. What are these intriguing structures?

The earth is made of atoms. In crystals they are arranged in simple arrays. Rubik's cube illustrates an array of cells, together filling space. Suppose each cell contained one atom. They would diffract beams of electrons or X-rays or neutrons producing patterns. The symmetries of the patterns are the same as the symmetry of the crystal. Two electron diffraction patterns are shown overleaf.

[xxii] Among quasicrystal researchers of course – details in part II.
[xxiii] see reference 1 on page 1.
[xxiv] Senechal, M., *Quasicrystals and geometry*, Cambridge, 1996

Simple cubes fill space (above) and produce diffraction patterns, (below) with regular symmetries and – in electron diffraction - beam spacings. The *Al* matrix in dual phase quasicrystalline Al_6Mn is quite similar to this.

0,0,0

Face shared dodecahedra cannot fill space. The faces are pentagonal. A 5-fold symmetry is implied by the following quasicrystal diffraction pattern below. Notice, in radial directions, the geometric series spacing, actually Fibonacci.

0,0,0

Diffraction patterns are formed when beams of electrons, X-rays or neutrons scatter off a solid. Each type of beam has special merits. Electron beams were used first because the specimens were microscopic while the electrons are easily focused by magnetic lenses. With larger specimens X-rays and neutrons provide extra information, but not necessarily about the important structural defects.

Quasicrystals are partly glass, partly crystal. They were discovered in 1982 but refereeing held up publication for two years. The figure above shows how difficult it is to fill space with a unit having dodecahedral symmetry and shared faces. The diffraction pattern from a quasicrystal having compatible symmetry with the cells is shown below it. Quasicrystals are not crystalline as follows from many reasons, one of which is their symmetry: the patterns are incompatible with space filling in conventional periodic arrays of atoms. Notice, moreover, the differences in the type of spacing between diffracted beams: in crystals the spacings are regular and arithmetic; in quasicrystals they are, typically, geometric. This geometric series is a special one called the Fibonacci series. Its terms are properly geometric, but each term is also the sum of the two preceding terms. This has special importance for double diffraction, when diffracted beams become sources for secondary diffraction. This is explained further in part II.

Nor are quasicrystals properly amorphous, or glassy, because the patterns are discrete. In amorphous materials cells are not aligned and the diffraction patterns consist in diffuse concentric rings having radii in regular (Bragg) sequences. In 1991, a new definition of crystals, based on discreteness of diffraction patterns, was agreed by the International Union of Crystallography[xxv].

[xxv] http://en.wikipedia.org/wiki/International_Union_of_Crystallography

Various types of glasses occur. The simplest is made from one type of atom, say silicon, where the deposition and cooling rate have been too rapid to allow the atoms to arrange in rows and layers. This material can be used for solar panels. Window glass is a little more complicated. Tightly bound $[SiO_4]^{++}$ tetrahedra join edge-to-edge with other tetrahedra in a more or less random way. The structure of quasicrystals has something in common with silica glass because they too have a single structural unit, joined at edges; but the quasicrystal units are, by this mechanism, directionally ordered. Illustrations will appear in part II. The bonding in the icosahedral alloy is not as strong as in $[SiO_4]^{++}$, so defects are more easily accommodated.

Now you can think of a quasicrystal as having one cell, made by a chemical driving force, with numerous structural defects; alternatively, you can think of it as having myriad different kinds of cell. Consider the simplest icosahedral cell first. It can be represented by a triadic golden rectangle. 'Golden' refers to the ratio of rectangular length to width. This is the ratio in a Fibonacci[xxvi] series $1, \tau, \tau^2, \tau^3...$, where the golden ratio, $\tau \equiv (1+\sqrt{5})/2$. By a special property of the series, each term is also the sum of the two preceding terms. In part II of this book several configurations of golden triads are illustrated. When they are attached to neighbouring triads by two edges, their orientations are aligned, and this allows the icosahedral symmetry and patterns to propagate.

If the rectangles were to touch randomly, you might have a truly glassy structure, as with silica, having diffraction patterns showing rings typical of disorder. This is not what is observed as you will see.

[xxvi] http://en.wikipedia.org/wiki/Fibonacci

After 1984, the new materials generated an industry of research. It was inspired by earlier mathematical speculations about how space can be filled using tiles compatible with 5-fold symmetries. In nature such symmetry is common, as in 5-petalled flowers, but in crystallography they are forbidden. Mackay[xxvii] had shown that the symmetry can be made with two tiles and he analyzed their properties. The physical scientists, inspired by this mathematics, spent their energy on juggling three-dimensional jig-saw puzzles to compare with their diffraction patterns.

They did not, however, notice that the original 1984 diffraction pattern[xxviii] of a rapidly quenched alloy, Al_6Mn, was not perfectly icosahedral as supposed; and went on to concentrate on the puzzle instead of the chemical driving force. The result was that, for many years, research into the dual foundation for the structure has been neglected: what is the chemical driving force for the structure and what, in detail, are its glassy defects?

The first discovery of quasicrystals was in a two phase material containing an Al matrix with dissolved Mn. The atomic concentration ratio less than 6:1 in the matrix; the second phase was Al_6Mn quasicrystal. By comparing the space occupied by atoms in the matrix from those in the quasicrystal, it is clear that the driving force is the formation of dense icosahedral cells, each a 'subcluster' of $Mn + 12\ Al$ atoms. The orientational symmetry of this unit is preserved when it shares edges with neighbouring subclusters. It can do this in such a way that clusters and superclusters of aligned icosahedra can in principle be formed. The clusters

---

[xxvii] Mackay, A.L., *Physica,* **114A**, 609-613 (1982).
[xxviii] Schechtman, D., Blech, I., and Gratias, D and Cahn, J.W., *Phys. Rev. Lett.,* **53**, 1951 (1984)

are readily seen in high-resolution electron microscope images[xxix], to be described in part II. They show also that the clusters are modified by defects at cluster centres and at fault lines between superclusters.

With the approximation of perfect superclusters, both the pattern and the intensities of diffracted beams are correctly calculated, proving the value of the extremely simple model: the units are logarithmically periodic, contrasting with the linear periodicity of regular crystals.

In summary, the following observations describe the history of quasicrystal research:
- It failed to identify the chemical driving force for the extraordinary structures[xxx].
- It was carried away by concerns of pure mathematics to the neglect of regular science.
- It was side tracked by puzzle building.
- It failed to notice that the original data is not icosahedral[xxxi] and to investigate pattern defects.
- It simulated understanding by excessively complicated methods for indexation of the pattern where simple inspection is complete, adequate and better. The method led to overcomplex computation.
- It failed to understand exceptions from Bragg diffraction[xxxii], including logarithmic orders, double diffraction and differences in interplanar spacing.

You can't get this litany into the journals for reasons already described. But the story implies a new condition for paradigm shifts in science. The shift in condition has

---

[xxix] See http://www.UHRL.net
[xxx] Appendix C in part II
[xxxi] Appendix D in part II
[xxxii] Appendix B in part II

35

become apparent since Kuhn's *Structure of scientific revolutions*[xxxiii].   Smolin's hopelessness is unnecessary. Open reviewing would be a first step.   But to break the network of petty power and self-interest, someone must upset the tables of the bankers.

The following model is an idealization of what is, experimentally, a quasiglassy structure. The model is used to explain all of the important features of the extraordinary diffraction patterns.   The complexity of the argument has been signposted by the mathematical conventions of lemma, proof and corollary.   However, when the meanings of signposts was not obvious or clear, it was sometimes preferable to place the sign *after* the demonstration, so as to point to where you have been, rather than where you are going. The flexibility was designed to best orient the reader for the next step.

[xxxiii] Kuhn, T.S., *The structure of scientific revolutions,* Chicago, 3rd ed., 1996

Quasicrystal or quasiglass?

The Structure of Quasicrystals

- *implied by the first explanation for fortuitous CBED[+],*
- *confirmed by optimum defocus HREM[++],*
- *with a single driving force,*
- *having structure factors for logarithmically periodic superclusters correctly calculated, and*
- *consistent with a new Compromise Spacing Effect*

+Convergent beam electron diffraction
++High-resolution electron microscopy

Antony J. Bourdillon

UHRL, P.O. Box 70001, CA 95170
December 2007, plus developments of November 2008,

The quasi-glassy structure of quasicrystals is explained:
The science lies in understanding the driving force
- rather than in constructing a jig-saw puzzle.
There is new physics because:
the diffraction does not follow Bragg's law,
and there are new effects.

Contents of part II

1.   Introduction II.

The discovery, twenty five years ago, of quasicrystals[1] having icosahedral symmetry, broke a fundamental tenet of crystallography and led to a new definition of crystals[2]. The new materials produce diffraction patterns that are too discrete for glasses and incompatible with crystals.

> *Crystals* contain arrays of atoms ordered periodically. These are revealed in diffraction patterns due to X-ray, electron or neutron beams. The patterns have the same symmetry as the crystals. These symmetries are described by the fourteen Bravais lattices that allow space to be filled. By simple mathematics it is possible to construct the atomic structure in real space from the diffraction maps in reciprocal space. The individual beams in the patterns are described by Bragg's law: for diffraction of radiation wavelength λ and integer order n, $n\lambda = 2d \sin\theta$ where d is the spacing between planes of atoms that cause the respective beam of Bragg diffraction, and θ is the Bragg angle, the angle between the plane and the incident beam. The diffraction is specular (mirror-like) between planes of atoms.

41

Quasicrystals. In Bragg diffraction, n is integral; in quasicrystals, *n* is generally logarithmic, but sometimes linear; i.e. in Bragg diffraction, angular spacings between diffracted beams are in arithmetic series; in quasicrystals they are generally in geometric series, or more particularly, the Fibonacci[6] series. Quasicrystals have icosahedral symmetry and this prevents crystalline space filling, since there is no corresponding Bravais lattice. Atomic arrangements are called 'aperiodic', because they are not arranged in regular linear arrays; though the supercluster structure that will be described is geometrically periodic. The diffraction pattern is sharp. The first problem is to describe how space is filled in a way consistent with the symmetry. The first step towards the solution is to identify the peaks in the diffraction patterns by the method called indexation.

2. Indexation of diffraction patterns

It is surprising, in retrospect, that up to 14 pages of abstruse mathematics[3,4] should be employed to index diffraction patterns incompletely, when the **complete indexation can be performed by simple inspection of the experimental patterns.**

-The indexation is facilitated by noticing that the icosahedron can be conveniently measured on orthogonal axes[3], rather than rhombohedral[4]. In the case of the 2-fold pattern, the indexation is further facilitated by noticing that the pattern is made of two parts – a Fibonacci spaced part and a Bragg part.

It is equally surprising, that the same mathematical effort should have been given to calculating the intensities of diffracted beams when the calculations are imperfect, having fudge (form) factors, from an incomplete indexation, ignoring double diffraction, dynamical effects, and atomic decoration, when
-The proper result can be accurately achieved by applying a one line, undergraduate level formula to any beam in the completely indexed pattern, using a unit structure of the greatest simplicity, modelled in a simple way. This unit structure is obvious, well-known, dense, and therefore chemically stable. The model will be described, and high-resolution electron microscope images confirm it.

You can shoot turkeys with an F18. One method simulates understanding; the other is more useful for discovering the structure, including defects. Physics should be made a simple as possible; not simpler; *not* as complicated as possible.

Quasicrystals, part II

Lemma 1. The diffraction patterns can be completely indexed by simple inspection

The structure of a regular icosahedron (appendix figure A3) contains 20 faces with 3-fold symmetry, 12 axes of 5-fold symmetry, and 30 sides with 2-fold symmetry. We assign axes [01τ] for the 5-fold; [111] for the 3-fold; and [001] for the 2-fold; where $\tau \equiv (1+\sqrt{5})/2$ is the golden ratio, so that in the Fibonacci sequence $\tau^{m-1} + \tau^m = \tau^{m+1}$, m being positive or negative integral. Considering first the decagonal pattern from the 5-fold axis, it is trivial to find the 10 vectors orthogonal to [01τ]. Radially, the indices of the Fibonacci series diffracted beams vary by the factor τ. Other beams can be indexed by linear interpolation. Figure 1 illustrates the main features of the Fibonacci series 5-fold pattern[1,5]. The symbols correspond with the indices detailed in table I. The ring of diffracted beams labelled C is especially bright on all three axial patterns, and will be used to determine the quasilattice parameter, as in appendix A. This fixes the scale of the unit structure and model.

The indices are represented in three equivalent forms in tables I-III, the decimal form being most convenient for the Fourier analysis contained in the structure factor calculation (equation 3 below): The calculations are conventional for centro-symmetric cells, but the cells are large, containing up to 250,000 atomic sites in supercluster order 3 (section 3).

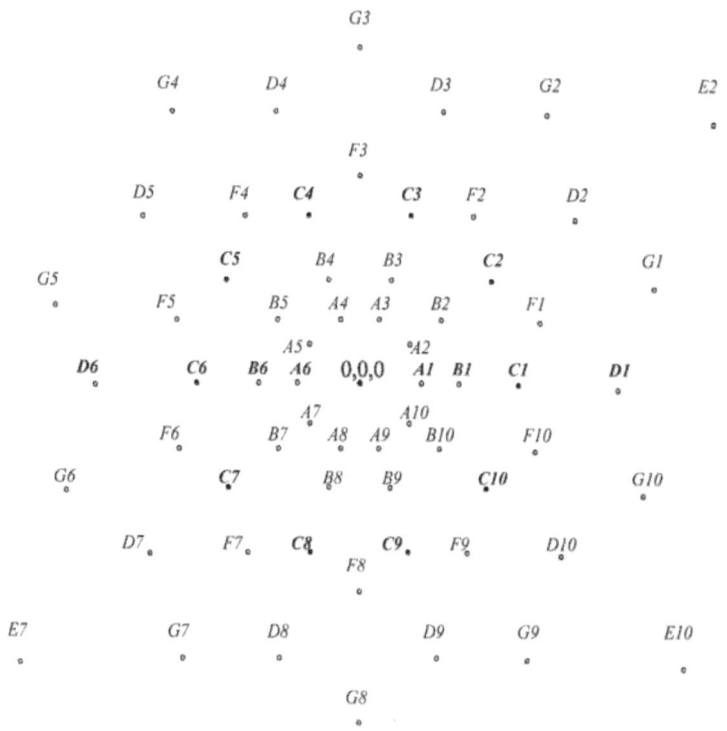

$$[0\underline{1}\tau]$$

Figure 1. Indexation of the 5-fold axial diffraction pattern in quasicrystalline *Al₆Mn* with symbols listed in the following table I, where the same indices are represented in power terms; in linear terms; and in decimals. The last are used in the structure factor calculations used to calculate beam intensities. Corresponding structure factors are given in the table, though their method of calculation will be described in section 4.

45

	h	k	l	h	k	l	h	k	l	Comment	factor**
										Indices for 5D axial pattern corresponding to figure	
A1	2/ttt*	0	0	4t-6	0	0	0.472136	0	0	C1 / tt	19.8
A2	1/tt	1/ttt	1/tttt	2-t	2t-3	5-3t	0.38196	0.236068	0.145898		"
A3	1/tttt	1/tt	1/ttt	5-3t	2-t	2t-3	0.145898	0.38196	0.236068		"
A4	-1/tttt	1/tt	1/ttt	3t-5	t-1	2-t	-0.145898	0.38196	0.236068		"
A5	-1/tt	1/ttt	1/tttt	t-2	2t-3	5-3t	-0.38196	0.236068	0.145898		"
A6	-2/ttt	0	0	6-4t	0	0	-0.472136	0	0		"
A7	-1/tt	-1/ttt	-1/tttt	t-2	3-2t	3t-5	-0.38196	-0.236068	-0.145898		"
A8	-1/tttt	-1/tt	-1/ttt	3t-5	t-2	3-2t	-0.145898	-0.38196	-0.236068		"
A9	1/tttt	-1/tt	-1/ttt	5-3t	1-t	t-2	0.145898	-0.38196	-0.236068		"
A10	1/tt	-1/ttt	-1/tttt	2-t	3-2t	3t-5	0.38196	-0.236068	-0.145898		"
B1	2/tt	0	0	4-2t	0	0	0.763932	0	0	C1 / t	298
B2	1/tt	1/tt	1/ttt	t-1	2-t	2t-3	0.618034	0.38196	0.236068		"
B3	1/ttt	1/t	1/tt	2t-3	t-1	2-t	0.236068	0.618034	0.38196		"
B4	-1/ttt	1/t	1/tt	3-2t	t-1	2-t	-0.236068	0.618034	0.38196		"
B5	-1/t	1/tt	1/tttt	1-t	2-t	2t-3	-0.618034	0.38196	0.236068		"
B6	-2/tt	0	0	2t-4	0	0	-0.763932	0	0		"
B7	-1/t	-1/tt	-1/ttt	1-t	t-2	3-2t	-0.618034	-0.38196	-0.236068		"
B8	-1/ttt	-1/t	-1/tt	3-2t	1-t	t-2	-0.236068	-0.618034	-0.38196		"
B9	1/ttt	-1/t	-1/tt	2t-3	1-t	t-2	0.236068	-0.618034	-0.38196		"
B10	1/t	-1/tt	-1/tttt	t-1	t-2	3-2t	0.618034	-0.38196	-0.236068		"
C1	2/t	0	0	2t-2	0	0	1.236068	0	0		573
C2	1	1/t	1/tt	1	t-1	2-t	1	0.618034	0.377211		"
C3	1/tt	1	1/t	2-t	1	t-1	0.377211	1	0.618034		"
C4	-1/tt	1	1/t	t-2	1	t-1	-0.377211	1	0.618034		"
C5	-1	1/t	1/tt	-1	t-1	2-t	-1	0.618034	0.377211		"
C6	-2/t	0	0	2-2t	0	0	-1.236068	0	0		"
C7	-1	-1/t	-1/tt	-1	1-t	t-2	-1	0.618034	0.377211		"
C8	-1/tt	-1	-1/t	t-2	-1	1-t	-0.377211	-1	-0.618034		"
C9	1/tt	-1	-1/t	2-t	-1	1-t	0.377211	-1	-0.618034		"
C10	1	-1/t	-1/tt	1	1-t	t-2	1	-0.618034	-0.377211		"
D1	2	0	0	2	0	0	2	0	0	C1 x t	546
D2	t	1	1/t	t	1	t-1	1.618034	1	0.618034		"
D3	1/t	t	1	t-1	t	1	0.618034	1.618034	1		"
D4	-1/t	t	1	1-t	t	1	-0.618034	1.618034	1		"
D5	-t	1	1/t	-t	1	t-1	-1.618034	1	0.618034		"
D6	-2	0	0	-2	0	0	-2	0	0		"
D7	-t	-1	-1/t	2-t	-1	1-t	-1.618034	-1	-0.618034		"
D8	-1/t	-t	-1	1-t	-t	-1	-0.618034	-1.618034	-1		"
D9	1/t	-t	-1	t-1	-t	-1	0.618034	-1.618034	-1		"
D10	t	-1	-1/t	t	-1	1-t	1.618034	-1	-0.618034		"
E1	2t	0	0	2t	0	0	3.236068	0	0	c1 tt	392
E2	tt	t	1	tt	t	-1	2.618034	1.618034	1		"
E3	1	tt	t	1	tt	t	1	2.618034	1.618034		"
E4	-1	tt	t	-1	tt	t	-1	2.618034	1.618034		"
E5	-tt	t	1	-tt	t	1	-2.618034	1.618034	1		"
E6	-2t	0	0	-2t	0	0	-3.236068	0	0		"
E7	-tt	-t	-1	-tt	-t	-1	-2.618034	-1.618034	-1		"
E8	-1	-tt	-t	-1	-tt	-t	-1	-2.618034	-1.618034		"
E9	1	-tt	-t	1	-tt	-t	1	-2.618034	-1.618034		"
E10	tt	-t	-1	tt	-t	-1	2.618034	-1.618034	-1		"
F1				3-t	2-t	2t-3	1.38196	0.38196	0.236068	C1+A3	75.7
F2				3t-4	1	t-1	0.854102	1	0.618034	C3+A1	"

continued

F3	0	2/t	2/tt	0	2t-2	4-2t	0	1.236068	0.763932	C2-D1+C2	75.7
F4				4-3t	1	t-1	0.854102	1	0.618034	A6+C4	"
F5				t-3	2-t	2t-3	-1.381966	0.381966	0.236068	C6+A4	"
F6				t-3	t-2	3-2t	-1.38196	-0.38196	-0.236068		"
F7				4-3t	-1	1-t	-0.854102	-1	-0.618034	etc.	"
F8	0	-2/t	-2/tt	0	2-2t	2t-4	0	-1.236068	-0.763932		"
F9				3t-4	-1	1-t	-0.854102	-1	-0.618034		"
F10				3-t	t-2	3-2t	1.381966	-0.381966	-0.236068		"
G1							2.236060	0.618034	0.38196	F1 x t	232
G2							1.381966	1.618034	1		"
G3	0	2	2/t	0	2	2t-2	0	2	1.236068	etc.	"
G4							1.38197	1.618034	1		"
G5							2.23607	0.618034	0.23607		"
G6							-2.236060	-0.618034	-0.38196		"
G7							-1.381966	-1.618034	-1		"
G8	0	-2	-2/tt	0	-2	2-2t	0	-2	-1.236068		"
G9							-1.38197	-1.618034	-1		"
G10							2.236060	0.618034	0.38196		"

* The golden mean is t=(1+sqrt(5))/2; 1/ttt symbolizes t to the power -3 etc.
**Structure factor: peak height x FWHM, for supercluster order 2 (in units of 1 thousand)

Table I. Indices and structure factors squared for 5-fold axial pattern. Notice that, as in the table footnote, $t \equiv \tau$, $.1/tt \equiv \tau^{-2}$, etc. The method for calculating structure factors will be described below in section 4. The indices are given in three equivalent forms: power terms, linear arithmetic and decimal. The last is for the structure factor calculations.

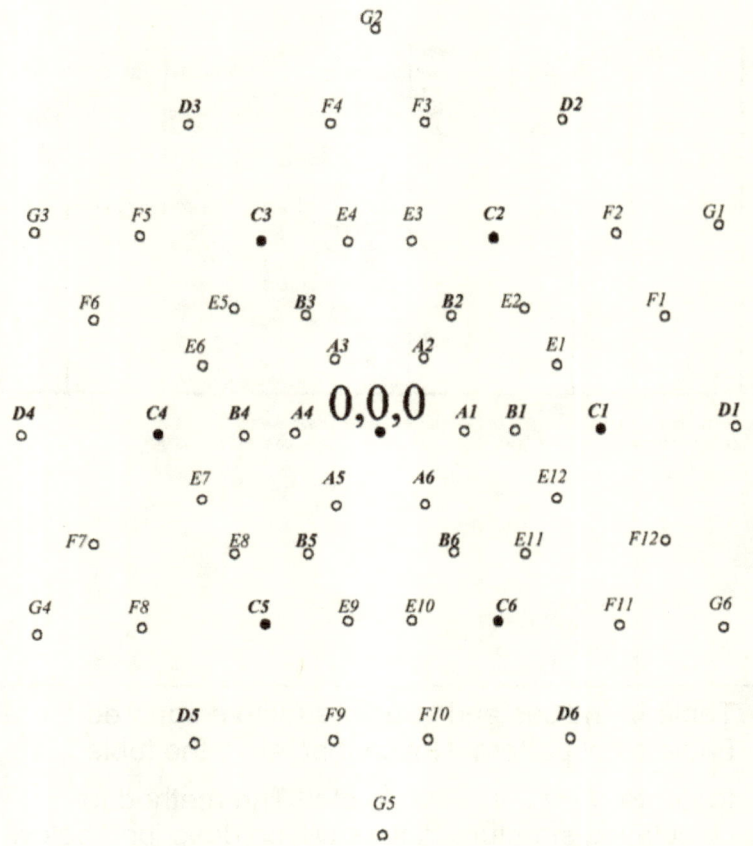

Figure 2. Indexation of the 3-fold axial diffraction pattern in *Al₆Mn* with symbols listed in the following table II, where the same indices are represented in power terms; in linear terms; and in decimals. The last were used in the structure factor calculations.

Table II. Indices and structure factors squared for the 3-fold axial pattern.

	h	k	l	h	k	l	h	k	l	comment	Str. factor
A1	-1/tt	1/ttt	1/tttt	t-2	2t-3	5-3t	-0.38196	0.236064	0.145893	C1 / tt	20.4
A2							-0.1458934	0.38196	-0.236064	C2 / tt	"
A3							0.2360643	0.145893	-0.38196	etc.	"
A4							0.38196	-0.236064	-0.145893		"
A5							0.1458934	-0.38196	0.236064		"
A6							-0.2360643	-0.145893	0.38196		"
B1	-1/t	1/tt	1/ttt	1-t	2-t	2t-3	-0.618034	0.381966	0.236064	C1 / t	300
B2							-0.2360643	0.618034	-0.381966	C2 / t	"
B3							0.381966	0.236064	-0.618034	etc.	"
B4							0.618034	-0.381966	-0.236064		"
B5							0.2360643	-0.618034	0.381966		"
B6							-0.381966	-0.236064	0.618034		"
C1	-1	1/t	1/tt	-1	t-1	2-t	-1	0.618034	0.38196		575 k
C2	-1/tt	1	-1/t	t-2	1	1-t	-0.38196	1	-0.618034		"
C3	1/t	1/tt	-1	t-1	2-t	-1	0.618034	0.38196	-1		"
C4	1	-1/t	-1/tt	1	1-t	t-2	1	-0.618034	-0.38196		"
C5	1/tt	-1	1/t	2-t	-1	t-1	0.38196	-1	0.618034		"
C6	-1/t	-1/tt	1	1-t	t-2	1	-0.618034	-0.38196	1		"
D1	-t	1	1/t	-t	1	t-1	-1.618034	1	0.618024	C1 x t	546
D2							-0.6180243	1.618034	-1	C2 x t	"
D3							1	0.618024	-1.618034	etc.	"
D4							1.618034	-1	-0.618024		"
D5							0.6180243	-1.618034	1		"
D6							-1	-0.618024	1.618034		"
E1							-0.7639274	0.763926	0	B1+A2	25.8
E2							-0.6180243	0.854098	-0.236073	A1+B2	"
E3							0	0.763927	-0.763926	A3+B2	"
E4							0.2360726	0.618024	-0.854098	B3+A2	"
E5							0.7639357	0	-0.763926	C4+B2	"
E6							0.8541066	-0.236074	-0.618024	C4+A2	"
E7							0.7639357	-0.763927	0	C4+A6	"
E8							0.618034	-0.854098	0.236074	C4+B6	"
E9							0	-0.763936	0.763927	C5+A1	"
E10							-0.236074	-0.618034	0.854098	C5+B1	"
E11							-0.7639274	0	0.763936	C6+A2	"
E12							-0.8540983	0.236074	0.618034	C6+B2	"
F1							-1.2360643	1.236068	0	C1+B2	201
F2							-1	1.381966	-0.38197	B1+C2	"
F3							0	1.236068	-1.236068	B3+C2	"
F4							0.3819697	1	-1.381966	C3+B2	"
F5							1.236068	0	-1.236064	B4+C3	"
F6							1.381966	-0.38197	-1	C4+B3	"
F7							1.2360643	-1.236068	0	C4+B5	"
F8							1	-1.381966	0.38197	B4+C5	"
F9							0	-1.236064	1.236068	C5+B6	"
F10							-0.3819697	-1	1.381966	C6+B5	"
F11							-1.236068	0	1.236064	C6+B1	"
F12							-1.381966	0.38197	1	C1+B6	"
G1	-(3-t)	t	-1/ttt	t-3	t	3-2t	-1.38196	1.618034	-0.23606	C1+C2	78.6
G2	1/ttt	3-t	-t	2t-3	3-t	-t	0.23606	1.38196	-1.618034	C2+C3	"
G3	t	-1/ttt	-(3-t)	t	3-2t	t-3	1.618034	-0.23606	-1.38196	etc.	"
G4	3-t	-t	1/ttt	3-t	-t	2t-3	1.38196	-1.618034	0.23606		"
G5	-1/ttt	-(3-t)	t	2t-3	t-3	t	-0.23606	-1.38196	1.618034		"
G6	-t	1/ttt	3-t	-t	2t-3	3-t	-1.618034	0.23606	1.38196		"

Structure Factor: peak height x FWHM (units 1 thousand) for supercluster order 2

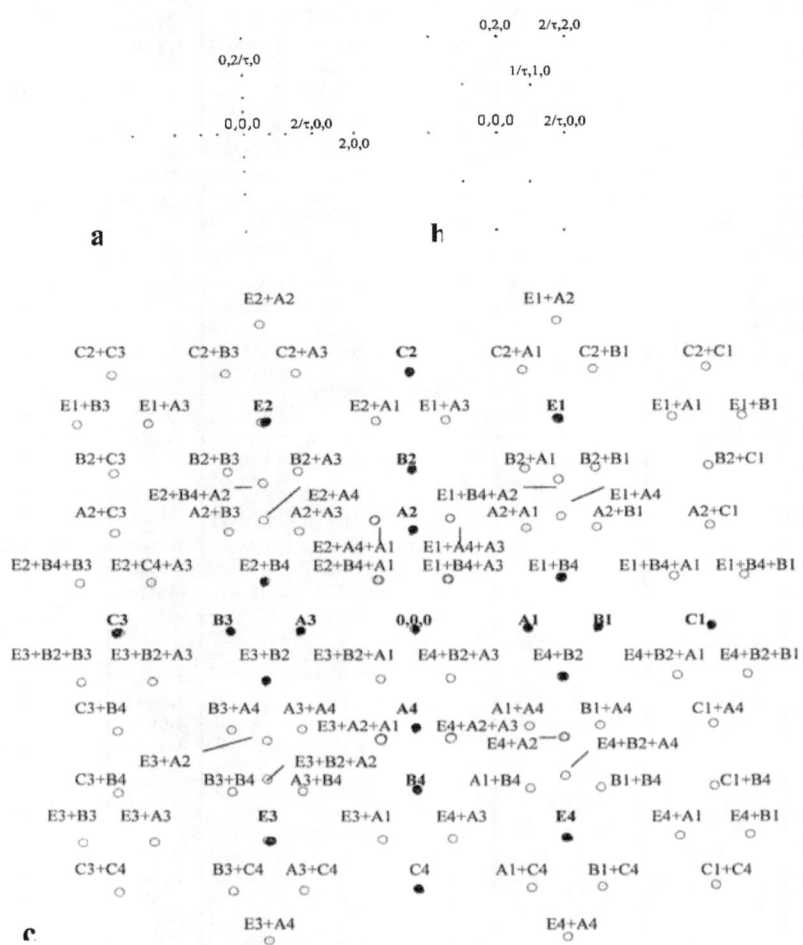

Figure 3.   2-fold diffraction pattern decomposed into
(a) Fibonacci series part and
(b) linear part. By repeating one pattern onto the other
   at bright spots – filled in
(c) the experimental pattern is reproduced and
   indexed with combined indices. The special
   double diffraction is described in section 4.

	Indices for 2D axial pattern										
	h	k	l	h	k	l	h	k	l	Comment	Factors**
A1	2/ttt*	0	0	4t-6	0	0	0.473126	0	0	C1 / tt	19.8 k
A2	0	2/ttt	0	0	4t-6	0	0	0.473126	0	etc.	"
A3	-2/ttt	0	0	6-4t	0	0	-0.473126	0	0		"
A4	0	-2/ttt	0	0	6-4t	0	0	-0.473126	0		"
B1	2/tt	0	0	4-2t	0	0	0.763932	0	0	C1 / t	298 k
B2	0	2/tt	0	0	4-2t	0	0	0.763932	0	etc.	"
B3	-2/tt	0	0	2t-4	0	0	-0.763932	0	0		"
B4	0	-2/tt	0	0	2t-4	0	0	-0.763932	0		"
C1	2/t	0	0	2t-2	0	0	1.236068	0	0		573 k
C2	0	2/t	0	0	2t-2	0	0	1.236068	0		"
C3	-2/t	0	0	2-2t	0	0	-1.236068	0	0		"
C4	0	-2/t	0	0	2-2t	0	0	-1.236068	0		"
D1	2	0	0	2	0	0	2	0	0	C1 x t	546 k
D2	0	2	0	0	2	0	0	2	0	etc.	"
D3	-2	0	0	-2	0	0	-2	0	0		"
D4	0	-2	0	0	-2	0	0	-2	0		"
E1	1/t	1	0	t-1	1	0	0.618034	1	0		955 k
E2	-1/t	1	0	1-t	1	0	-0.618034	1	0		"
E3	-1/t	-1	0	1-t	-1	0	-0.618034	-1	0		"
E4	1/t	-1	0	t-1	-1	0	0.618034	-1	0		"
F1	2/t	2	0	2t-2	2	0	1.236068	2	0	2E1,C1+D2,C2+D1	234 k
F2	-2/t	2	0	2-2t	2	0	-1.236068	2	0	2E2,C3+D2,C2+D3	"
F3	-2/t	-2	0	2-2t	-2	0	-1.236068	-2	0	2E3,C3+D4,C4+D1	"
F4	2/t	-2	0	2t-2	-2	0	1.236068	-2	0	2E4,C1+D4,C4+D1	"
F5	3/t	1	0	3t-3	1	0	1.854102	1	0	E1+C1	132 k
F6	-3/t	1	0	3-3t	1	0	-1.854102	1	0	E2+C3	"
F7	-3/t	-1	0	3-3t	-1	0	-1.854102	-1	0	E3+C3	"
F8	3/t	-1	0	3t-3	-1	0	1.854102	-1	0	E4+C1	"
E1+C4	1/t	-1/ttt	0	t-1	3-2t	0	0.618034	-0.23607	0		545 k***
E2+C4	-1/t	-1/ttt	0	1-t	3-2t	0	-0.618034	-0.23607	0		"
E3+B2	-1/t	1/ttt	0	1-t	2t-3	0	-0.618034	0.23607	0		"
E4+B2	1/t	1/ttt	0	t-1	2t-3	0	0.618034	0.23607	0		"

* t=(1+sqrt(5))/2 is the golden number. 1/ttt rpresents t tothepower -3, etc.
**Structure Factors: peak height x FWHM (units of 1 thousand) for supercluster order 1
***An example of a complex beam that is correctly predicted bright; cf weaker beams also predicted
   as follows:

Table IIIa. Indices of 2-fold diffraction pattern and calculated structure factors squared.

Table IIIb. Indices of beams in positive quadrant of 2-fold axial pattern and structure factors squared,
Bottom row shows forbidden line, the first diagonal in a 2-fold pattern rotated by 90°.   (overleaf).

Diffracted beams in the positive quadrant of 2D				experimental
	h	k	l	Factor*	rank**
	0	0	0		10
C1	1.236068	0	0	573 k	9
A1	0.473126	0	0	19.8 k	6
B1	0.763932	0	0	298 k	7
D1	2	0	0	546 k	8
E1	0.618034	1	0	955 k	9
F1	1.236068	2	0	234 k	5
F5	1.854102	1	0	132 k	5
E1+B4+B1	0.618034	0.236068	0	89.3	3
E1+B4+A1	1.09116	0.236068	0	57	3
E1+B4	0.618034	0.236068	0	545 K	8
E1+B4+A3	0.144908	0.236068	0	37	close to 0-order
A2+C1	1.236068	0.473126	0	4.3 k	4
A2+B1	0.763932	0.473126	0	2.6 k	4
A2+A1	0.473126	0.473126	0	231	3
E1+A4	0.473126	-0.473126	0	231	3
E1+A4+A3	0.144908	0.526874	0	1.4 K	3
E1+B4+A2	0.618034	0.709194	0	681	1
B2+C1	1.236068	0.763932	0	75 k	5
B2+B1	0.763932	0.763932	0	25 k	4
B2+A1	0.473126	0.763932	0	255	3
E1+B1	1.381966	1	0	327 k	5
E1+A1	1.09116	1	0	1.2 k	3
E1+A3	0.144908	1	0	782	5
C2+C1	1.236068	1.236068	0	201 k	5
C2+B1	0.763932	1.236068	0	70 k	4
C2+A1	0.473126	1.236068	0	1.9 k	2
E1+A2	0.618034	1.473126	0	15.4 k	5
Rotated E1***	1	0.618034	0	17	0

*Structure factors calculated for supercluster order 2
** Interpreted from photographs in ref . 1, Schechtman et al., with
            scales from 10 (zero order) to 0 (forbidden)
***Not data, but calculated as an example of a forbidden line

The structure factors were calculated for the various diffracted beams to give a first approximation for their intensities, ignoring for simplicity and reliability, scattering angle and reflection sphere effects. The factors match, without error, the rankings interpreted from published diffraction data[1]. Further details will be given below after the description of the model in section 3.

Figure 2 and table II show indices and calculated structure factors for the 3-fold symmetric axial patterns. As in the 5-fold pattern, the spacings between diffracted beams all follow Fibonacci series.

The 2-fold pattern, illustrated in figure 3 and tables IIIa and IIIb, is more complicated. The pattern consists of two parts: a vertical-horizontal cross pattern spaced in Fibonacci series, while diagonals are spaced on a regular linear Bragg lattice. When one pattern is placed over the bright spots of the second, the experimental pattern is reproduced. When this pattern is indexed and structure factors calculated, the strengths of the individual beams match the rankings in experimental data without error. The result conclusively supports the model, which will be described next. The superposition follows double diffraction, special in quasicrystals, as described below in section 4.

Lemma 2. The diffraction has two parts: Fibonacci series and Bragg

The forbidden line shown in table IIIb is the first diagonal diffraction spot on the anomalous experimental D_2^{525} pattern (the 2-fold pattern[1] on the 5-2-5 stereographic projection) illustrated in appendix D. The calculations therefore support either a data representation error in the original data of Schechtman et al[1], or, alternatively, a genuine defect in icosahedral symmetry. Until the pattern is known, it is pointless to examine possible divergence from icosahedral symmetry. Such divergences are expected due to both directional solidification and to glassy defects.

Proof 1. The intensities of diffracted beams are correctly calculated

3. Model

Subclusters, clusters and superclusters

>*This model differs from others'* [7.9] *by giving primacy to the chemically stable subcluster. This is the structural determinant. Defects occur at the cluster and supercluster levels and these reflect back on the subclusters. Regular superclusters are used approximately to calculate diffracted intensities and to describe other properties..*

Lemma 3. The icosahedral subcluster is dense and therefore chemically stable.

Any solution for the structure of quasicrystalline Al_6Mn must contain an answer to the following consideration. The icosahedral second phase grows in a face centred cubic matrix that is one of the most common structures in metals. In this matrix, *Mn* is depleted[5]. Why does the segregational increase in *Mn* concentration in the quasicrystalline second phase, radically change the structure ? The formation of icosahedral subclusters is the most obvious reason. Figure 4a shows an icosahedral subcluster consisting of a central *Mn* atom bound tightly, with high coordination, to twelve *Al* atoms. The diameter of *Mn* is 12% smaller than *Al*. In the Al matrix, *Mn* solute atoms have space to 'rattle' in the fcc solvent (appendix C). In the quasicrystalline second phase, bonding causes a contraction of the *Al* diameters as shown by the measured quasilattice parameter. In consequence, the volume of this subcluster is 17% smaller than the volume occupied by a similar set of atoms in the face centred cubic matrix. This reduction in volume in the quasicrystalline phase corresponds to a reduction in enthalpy that stabilizes the subclusters and is the driving force for the phase.

> *Two conditions for the formation of these subclusters are evident:*
> *-a chemical ratio of 6:1 between solvent and solute atoms;*
> *-a ratio of their radii* $R_{solute}/R_{solvent} = \sqrt{\tau^2 + 1} - 1$.

Corollary 1. The driving force is the low enthalpy of the dense subcluster

The subcluster can be represented by a triad of golden rectangles with *Al* at each vertex.   Each triad can represent either the icosahedron or its reciprocal Platonic solid, the dodecahedron[6]. These solids have the same point group symmetry. The triads join in either quad concave or quad planar orientations by two triply shared vertex *Al* atoms and by 4 doubly shared *Al* atoms, so as to include four edges (figure 4b).   Six overlapping concave quads form an icosahedral cluster of twelve subclusters (figure 4c). The icosahedral cluster is itself dense, with twenty triple vertices (each shared by three subclusters); twelve dangling sites at the centre; and the remaining outer atoms shared with neighbouring clusters.

This gives the basis for the icosahedral symmetry in the observed electron diffraction patterns[1]. Any theory that does not recognize a driving force allows for an economy here in relevant citations. To avoid negative controversy, we use a clean sweep with a new brush.

The edge length of the cluster is τ^2 times that of the subcluster. The relationship between subcluster and cluster can be extended to superclusters of increasing order; all with perfect icosahedral symmetry; all identically oriented; and each represented by figure 4c. Likewise these structures can all be represented, at appropriate scales, by golden triads. Differences at the respective centres are described below in association with defects.

A convenient feature of the model in figure 4c is easy recognition of coordinates for computing purposes. This recognition facilitated the calculations listed in tables IIIb.

a

b

c

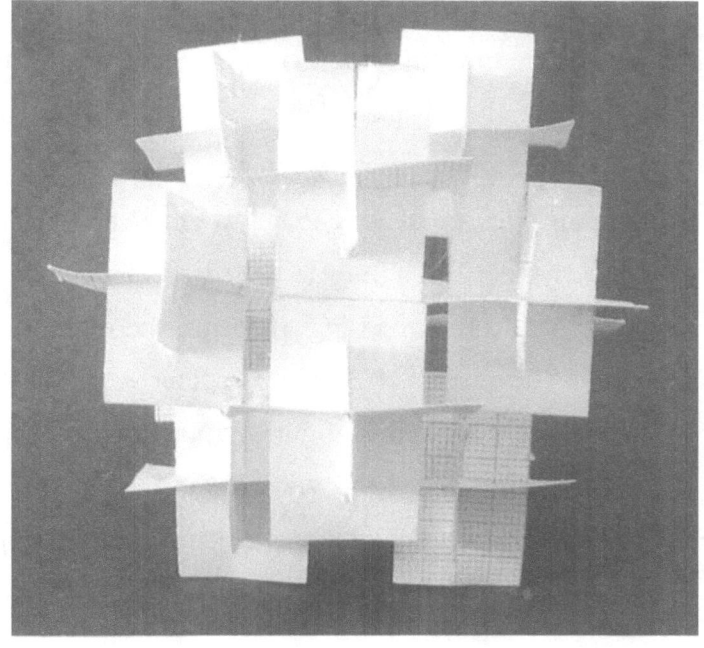

Figure 4. (a) Icosahedral subcluster, of 12 *Al* atoms surrounding a central *Mn* atom, represented by triadic golden rectangles in
(b) either concave (right) or planar (left) orientations that can combine to form
(c) an icosahedral cluster, viewed parallel to the [001] axis. Planes of *Al* atoms can be readily identified within clusters and across them. They serve as quasi Bragg planes that include varied spacings of $\tau/2$, 1/2, and $1/2\tau$ where τ is the length of a subcluster. Compare inferior space filling of dodecahedra on p. 31.

Glassy defects in clusters

As has been previously observed[7], the centre of the cluster is problematic. It has been described as a hole; in fact it is the densest part of the structure owing to dangling *Al* atoms: their overlapping sites are dense enough to force vacancies. In a cluster (figure 4c) the 12 dangling icosahedral sites at the centre allow space for only three opposing *Al* atoms, the other nine sites being vacant. The proposed centre of a cluster is illustrated by the schematic cross-section in figure 5a and of two superclusters in figure 5b. This illustrates overlaps and ingrowths. The latter are suggested by high-resolution electron microscope (HREM) images, recorded at the optimal defocus condition[8], as below.

Cluster centre

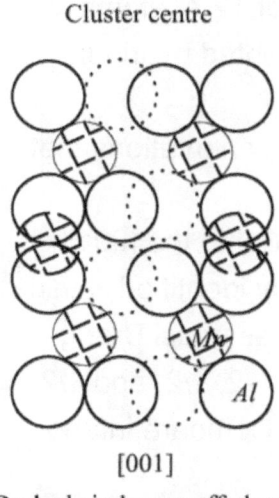

[001]

Dashed circles are off plane
Dotted circles are vacancies

Figure 5a. Section through centre of cluster viewed from the [001] axis. Crossed circles represent *Mn*; open *Al*. Dotted circles show the likely positions of vacancies where space is not available for site occupation. Dashed circles on *Mn* atoms are off the sectioned plane and do not overlap neighbouring atoms; they are added to illustrate completeness of the cluster.

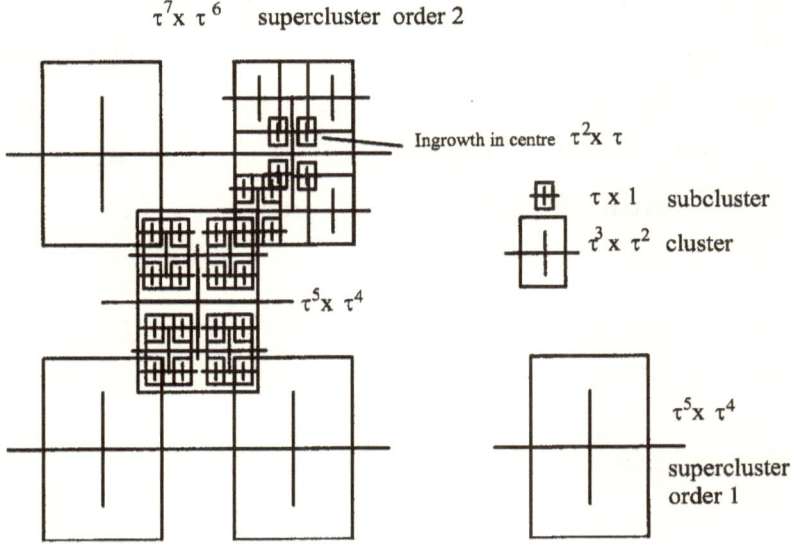

Figure 5b. Cross-section of superclusters, represented by golden rectangles due to the superclusters of orders 2 and 1, and in decreasing size, the sections of constituent clusters and subclusters. The figure indicates how subcluster ingrowths from clusters can fill the central space of the cluster by overlapping. Near its centre, the gap between subclusters is, conveniently, one *Al* atomic diameter. HREM data, described below, provide evidence for such ingrowths where they appear as hull structures. Equivalent outgrowths are also evident on fault lines as described below. The figure illustrates a section containing four orders of clusters within clusters.

At this optimum defocus, darkening of the negative (bright in the print) corresponds to high atomic density. Such defects in a structure conceived from a single unit is an economical, alternative description to multiple varieties of cells[7]. It is economical because there is one driving force; the varieties correspond to defects.

Lemma 4. Space is filled by icosahedral substructures plus defects.

High-resolution electron microscopy (HREM)

Our figure 6 shows a sectional view of an icosahedral supercluster order 1 that correlates identically with features itemized in figure 5. Each cluster contains 12 subclusters arranged icosahedrally. Viewed from the 5-fold axis, the subclusters are arranged decagonally in the clusters. Each filled black circle represents a subcluster and corresponds, in reverse, to circular white contrast in the HREM image (figure 7a). A complete supercluster order 1 consists of 12 clusters arranged icosahedrally, but in figure 6a a plane of five clusters are selected. The section omits one (lower) half of the supercluster as well at the opposite (upper) apical subcluster. The section contains whole clusters, but only half a supercluster. More complete views of atoms in the section will beexamined later (figure 9b). Meanwhile, the arrangement in the section is made from five clusters, each containing ten subclusters. The clusters model the most common structure in figure 7a.

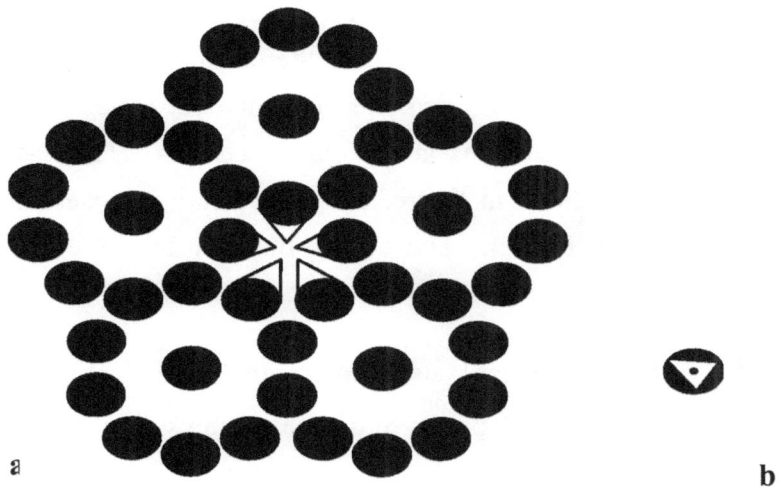

a   b

Figure 6.   (a) 5-fold axial view of a section of a supercluster order 1, made up of a pentagonal plane of clusters and omitting one (lower) half of the supercluster and the opposite apical cluster. Each filled circle represents a subcluster, consisting of a view of an icosahedral (decagonal in this orientation) arrangement of subclusters, each containing $Mn+12$ Al atoms. The subclusters are generally icosahedral, but they include vacancies. Our figure closely matches one of the structures observed in HREM (figure 7).
(b) Typical representation of cluster centre showing 3-fold symmetry due to 3-fold Al occupation of sites at cluster centres, with 9 vacancies. This 3-fold centre is evident in all clusters in figure 7a.

Whereas the clusters are decagonal, their centres show 3-fold symmetry. The 3-fold occupation of the cluster centres has been described above. The symmetry can, therefore, immediately be recognized as the three occupied sites and nine vacancies that are allowed by space restrictions at the cluster centres. This 3-fold symmetry is represented in figure, 6b. This detail is observed at the centre of *every* cluster. The observation confirms the triadic model with inclusive defects.

The HREM data of Bursill and Peng[8], notably their figure 2a (reproduced with permission in figure 7a), shows the following general and detailed features:

- Structures (corresponding dimensionally to our subclusters) that lie on long rows arranged pentagonally.
- The row symmetry derives from the HREM image being on the 5-fold axis, S_{10}, in the diffraction pattern.
- Multiple circular structures, arranged decagonally, that correspond to our clusters.
- Three specific features:
  - A (pentagonal) section of a supercluster order 1 at the lower right of their figure;
  - A close pack of six clusters at the upper left of their figure; and
  - A fault line that runs 15 degrees east of vertical north and parallel to one of the rows of subclusters.

- A detailed 3-fold structure at the centre of each cluster.
- A hull shaped structure (pointed at one end and oval at the other) that occurs with pentagonal symmetry at the centre of the pentagonal supercluster section, and
- Along the fault line with many occurrences.

All of these features correlate with our model. In particular, each discrete circle represents a relatively heavy *Mn* atom surrounded by its subcluster of *Al* atoms.

Notice a difference between figures 4c and 7b. In the former the triads represent subclusters; in the latter clusters. Figure 7b models a glassy arrangement of clusters. The model is not simply one of aligned icosahedra; it is infinitely extensive through superclusters of any order. Notice secondly that it is debatable whether multislice simulations can improve on the original data in this case. The image acquired at optimum defocus is clear in itself; simulations imply artifacts and can be manipulated. Simulations are generally used for interpretation of *crystal* images, off optimum defocus, when resolution is higher. However, multislice simulation programs generally assume dynamical diffraction and are not adapted to the peculiarities of quasicrystal diffraction.

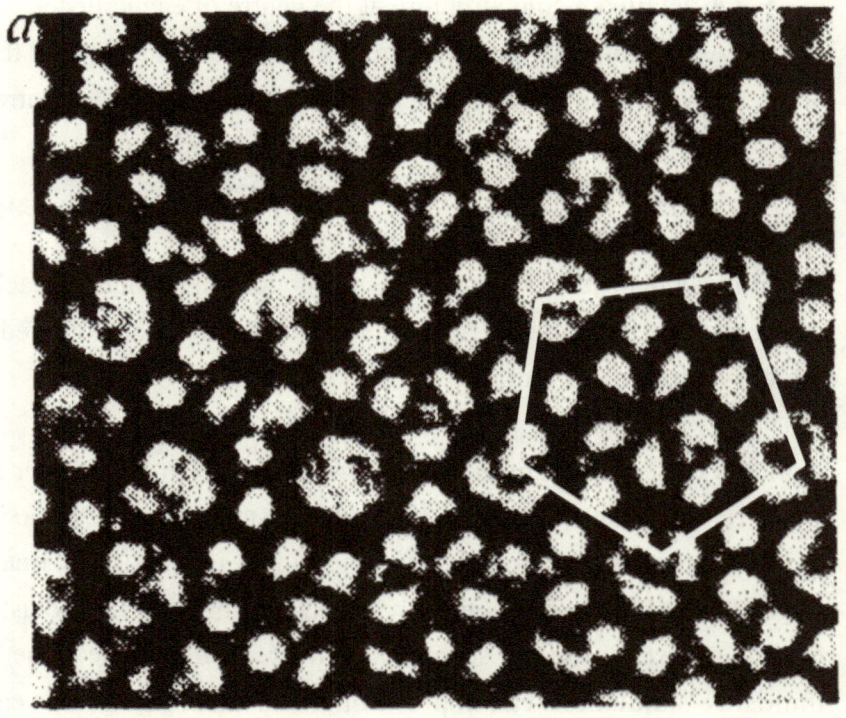

Figure 7a. HREM of Al_6Mn (ref. 8) recorded at optimum defocus. 3-fold centres (as in fig.6b) of regular decagons mark the cluster centres. Note the sections of supercluster order 1 outlined with a pentagon that connects cluster centres. At the centre of this supercluster order 1 are 5 hull shapes that are interpreted as ingrowths. Image reproduced with permissions from L.A.Bursill, J.L. Peng and from Nature, permission #2047130461239.

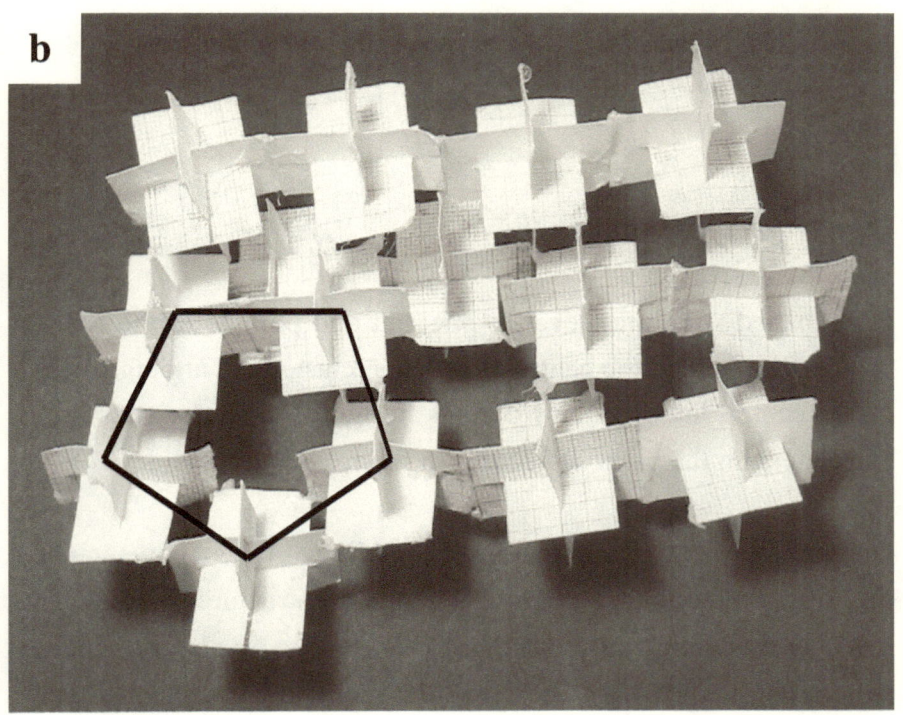

Figure 7b. In mirror image of figure 7a, skeletal structures representing icosahedral clusters, match the HREM image. Each corner of the golden rectangular triads locates an icosahedral subcluster containing central *Mn* + 12 *Al* atoms. Note the corresponding sections of supercluster order 1 outlined in each image with a pentagon that connects cluster centres.

Proof 2. The supercluster order 1, clusters and subclusters are observed.

Ingrowths and outgrowths

Hull shaped structures, oval at one end and pointed at the other, can be seen both at the centre of the supercluster and at outgrowths on the central fault line on 1 o'clock in figure 7a. These structures correlate with the proposed details in figures 5 and 6.

Our method contrasts with puzzle building. When you include ingrowths, outgrowths, vacancies etc into the structure, you begin a menagerie of atomic arrangements[9, 10], cells etc. We have a unique structure as driving force while including ingrowths and outgrowths as glassy defects.

Proof 3. Corresponding defects are observed.

4.   Diffraction

Convergent beam electron diffraction (CBED)

Consider how this model explains the salient features of electron diffraction. Firstly, all of the icosahedral symmetries must occur, including those in the 5-fold (C_{5v}, using Schoenflies notation – or S_{10} in the three dimensional model with inversion symmetry), 3-fold (C_{3v}, or S_6) and 2-fold (D_{2h}) axial patterns[1]. Secondly, consider the relative intensities and lattice parameter. In electron diffraction from thin foils[5], the dominant feature is the circle of brightly diffracted beams that occur at about the same radius in all three axial patterns just mentioned. This suggests a quasicell with approximate sphericity, as in figure 4a, but contrasting with the dual oblate and prolate rhombohedra proposed by others[11], or with multitudinous variants having unspecified chemistry[9,10]. The C_{5v} pattern, axial $[0\underline{1}\,\tau\,]$, and C_{3v} pattern, axial $[111]$, both contain radial arms with diffracted beams spaced in geometric series with Fibonacci power law scattering angles, $\Theta \propto \tau^m$, m integral. On the radial arms, the diffracted beams intersect concentric circles that we name A, B, C, D, beginning near the centre. The third, C, is the brightest[1.5], especially in thin foils that approximate kinematic scattering[13].

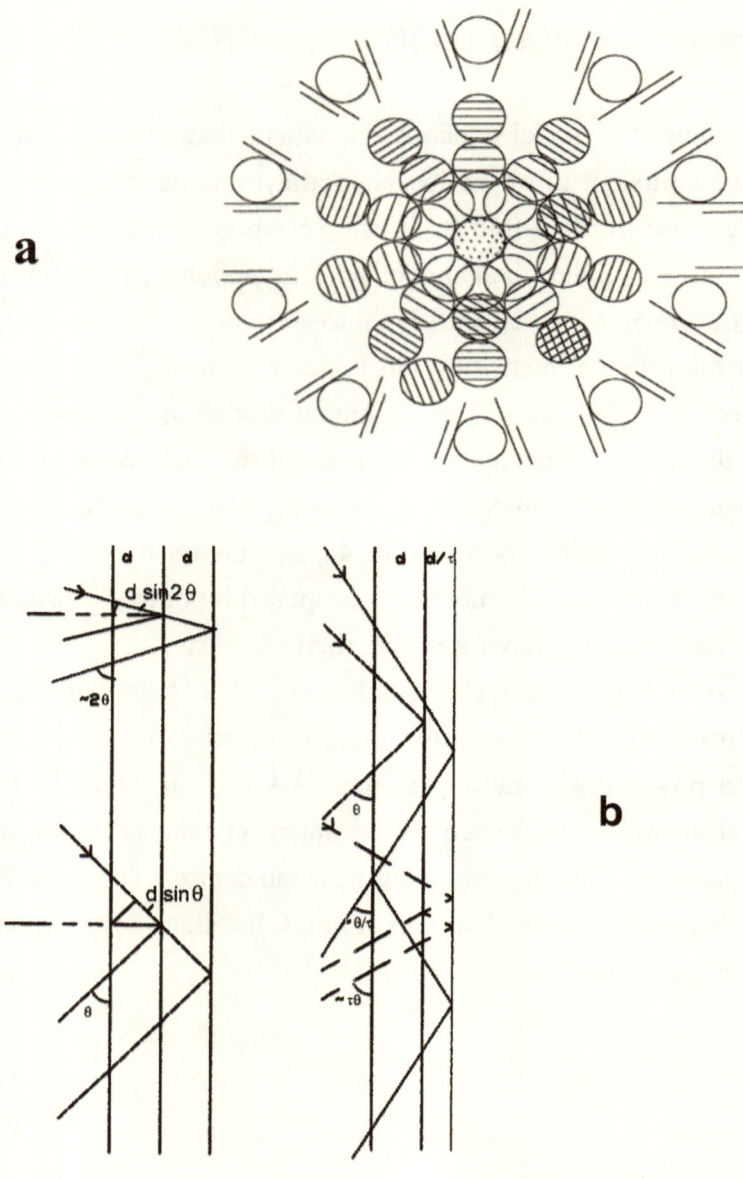

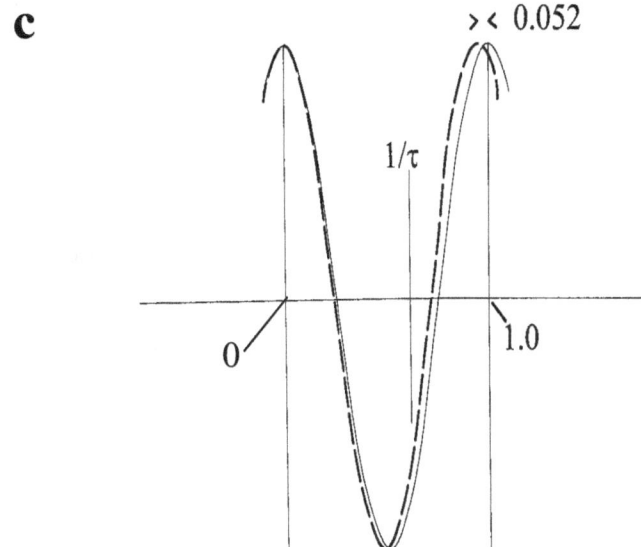

c

$1/\tau$

$> < \ 0.052$

0

1.0

Figure 8.

(a) Schematic of a fortuitous CBED pattern.   Lines oblique to the tangents are due to double diffraction.

(b) Illustration of diffraction from regular crystals (left): in schematic with small angles, $n\lambda \approx 2dn\theta$. In quasicrystals (right) $\lambda \approx 2d\tau^{-n}\theta$

(c) The new **Compromise Spacing effect** causes diffraction maxima to occur at corrected Bragg conditions, $\sin(\theta) = n\lambda / 2d'$ where the corrected $d'=d\,/\,1.0518$ at maximum diffracted intensity is smaller than the real quasicrystal interplanar spacing d.

(d) See next page.

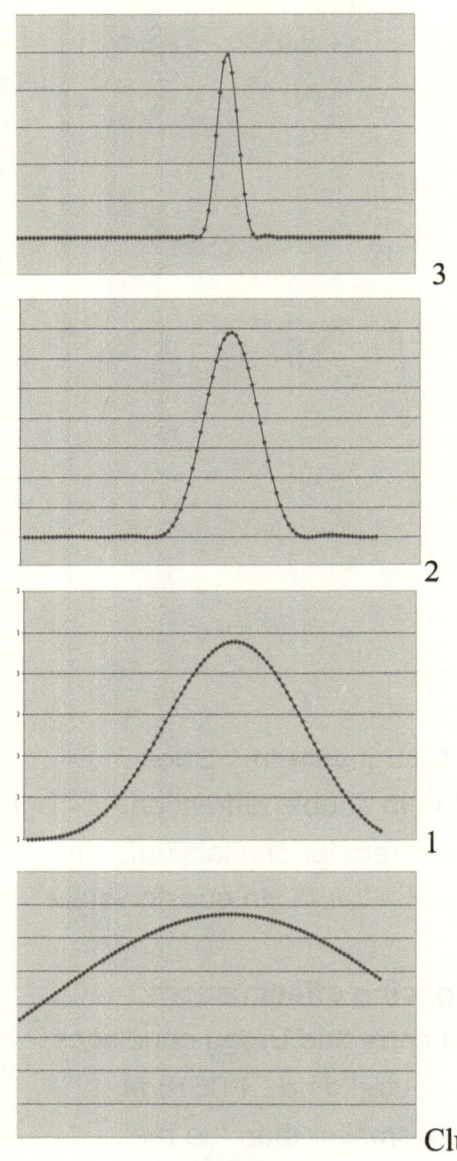

3

2

1

Cluster

0   0.947
Fractional divergence from Bragg angle at origin

Figure 8(d). Calculated $(2/\tau,0,0)$ 'structure factors' with changing diffraction angle.   In quasicrystals, diffraction peaks, are shifted from their corresponding positions (origin of abscissae) predicted by Bragg's law for crystals. Factors were calculated, in order from top, for: supercluster order 3; order 2; order 1 and cluster. In quasicrystals, the Compromise Spacing Effect shifts the calculated Bragg angle by 5.3%. In consequence the measured interplanar spacing is increased by 5-6% from the spacing previously 'measured' by supposing Bragg diffraction.

Figure 8a is a schematic representation of a fortuitous convergent beam diffraction pattern[5] that demonstrates important features in the unusual power law diffraction. The diffracted beams contain interference fringes due to a horizontal fissure roughly parallel to the plane of the foil. Their spatial frequency is proportional to the scattering angle. The convergent, zero-order beam shows spots due to the reverse of the intersecting tangential lines in the diffracted beams. In heavily exposed images, the interference was observed also in Kikuchi lines. These features can all be reproduced in crystals[12]. It is surprising, however, that the interference occurs not only between the zero order and diffracted beams, but also between diffracted beams - giving rise to oblique line orientations in the interference. The oblique lines are at large angles, close to normal to the primary interference with zero order. Scattering between diffracted beams can be understood with dynamical diffraction theory, but why does diffraction not proceed at angles $2\Theta, 3\Theta...$ etc?

Lemma 6. The Fibonacci series is filtered by alternating Fibonacci interplanar spacings.

In regular crystals, phase relationships force this linear series; in quasicrystals by contrast, alternating long and short interplanar spacings cause more complicated relationships between scattered beams that interfere constructively (figure 8b). There is a filtering effect in the constructive interference of waves scattered by atoms in different Bragg planes that are separated by different spacings in Fibonacci series. The necessary phase relationships, due to scatterings from the different interplanar spacings, can be described by the product (appendix B):

$$e^{2\pi i d/(2\lambda)} . e^{2\pi i d/(2\lambda\tau)} = e^{2\pi i d t/(2\lambda)} , \tag{1}$$

so that waves scattered sequentially between smaller spacings, on the left side of the equation, can superpose constructively onto waves scattered from larger spacings, on the right side. Similar relationships hold for other spacings in the Fibonacci sequence. This understanding will be further employed in our analysis of the D_{2h} pattern with its mixture of Fibonacci series and linear patterns. Beyond these differences, as described, we assume that the general features of Bragg scattering from a lattice occur similarly in quasicrystals. This diffraction in the quasicrystal is consistent with the particular structure factor calculations to be described below

Lemma 5. Double diffraction does occur, but only at near normal angles.

Fibonacci series diffraction

Notice that the Fibonacci power series implies a modification to Bragg's law for diffraction from crystals, $n\lambda = 2d\sin\theta$, n for diffracted order *n*, wavelength λ, interplanar spacing *d* and Bragg angle θ. In quasicrystals, *n* is restricted to 0 or 1: higher orders are generally (with exceptions) forbidden, including lines in higher order Laue zones (HOLTZ lines), as observed[5]. Since angles are small in high energy electron diffraction, we can write, as an approximation, for the Fibonacci series spacings d_m and scattering angle $\Theta \approx 2\theta$, $\lambda \approx d'\tau^{-m}\Theta$, where d' is the quasilattice parameter corrected for the Compromise Spacing effect.

This new effect constitutes a yet further difference between Bragg diffraction from crystals and diffraction from quasicrystals. The difference is illustrated in figures 8c and 8d and appeared universally in the structure factor calculations to be described. The difference results in a correction to the former measurement of interplanar spacing and results from series of unequal spacings. The quasi structure factor (adapted from equation 3 below) maximizes at angles (or wavelengths) smaller than expected (or conversely at inteplanar spacings larger than previously predicted) causing a correction factor in the measurement of the spacings. For a centro-symmetric system, after differentiating the cosine on three planes spaced by Fibonacci relationships, the amplitude maximizes when:

$$\frac{x}{\tau}\sin\left(\frac{2\pi + x}{\tau}\right) = x\sin(x)$$

(2)

as illustrated in figure 8c, where *x* is the correction factor for a measured spacing *d'=dx*. Owing to second order effects and

75

truncation limitations in calculations, the correction varied by 1.056±0.001 when calculated for supercluster order 2. The correction is critical for the space allowed for *Al* atoms in the subcluster structure. In brief summary, the bad news is that the atoms are too large – they don't fit. The good news is that the cell is larger than expected. The result is that the atoms do fit after all.

Lemma 7 and corollary 2 (consistently). The measured interplanar spacings are larger in Fibonacci series diffraction than in coresonding Bragg diffraction.

In icosahedral Al_6Mn, the reciprocal quasi-lattice is mostly Fibonacci. The condition for electron diffraction, with scattering vectors $K_I = k_{incident} - k_{scattered}$, can be described in the normal way[13] by $K_i = g_{i,}$ at reciprocal quasilattice vectors g_i . The basis in reciprocal space is the unit cube, the simplest possible basis, where the unit is the inverse of the side length of the subcluster. Simulations for this D_{2h} pattern have previously failed[11] or had varied results[9,10]. However, with our model and the following structure factor calculations, the pattern is now defined. Figure 9a is useful for understanding the necessity for the two types of diffraction pattern on the D_{2h} axis. The figure shows a view, from [001] of a supercluster order 1, of atoms including interatomic bonds. The diagonal angle is $\tan^{-1}(\tau)$ from the horizontal. The supercluster exhibits a variety of interplanar spacings between planes through subcluster centres that are normal to the diagonals shown. The dominating Fibonacci series spacings are $\tau / \sqrt{\tau^2 + 1}$

short and $\tau^2/\sqrt{\tau^2+1}$ long. Diffraction from these planes is not observed and their structure factors are calculated to be small. In fact, the observed linear diagonal pattern has the interplanar spacing d_d , equal to the semi-diagonals of the golden rectangle in the subcluster. It is clear therefore that the diagonal pattern arises from intra-subcluster distances. The diagonals, being due to the intra-subclustral scattering, have spacings that are linear, not Fibonacci. The calculations corroborate this view.

Returning to the general model, an examination of neighbouring sites at the centres of triadic rectangles, shows that they lie on a lattice $a+b\tau$, where a and b are any integers. Many sites are vacant where overlapping of quasicells prevents occupancy. Centres of neighbouring edge-vertex-edge (EVE) bonded triads are separated by the subcluster length τ . The scaling factor for this distance is given by identifying $\tau/2$ with the measured quasilattice parameter d_0 on the bright ring c_0 for reasons summarized in tabulation form in appendix A. The measured value[5,7,14] is, before correction, $d_0=$ 0.206nm. The length of the subcluster (or golden rectangle) becomes 0.412 nm, and the side length (width) is 0.254 nm. When these numbers are corrected by the factor 1.056, due to the Compromise Spacing effect described earlier, they translate to 0.435 nm long and 0.268 nm wide. This last dimension is 5% shorter than the normal diameter of *Al* atoms in the fcc metals (0.286 nm). The shortness matches the large coordination in quasicrystals, reaching 15 at triple points. The data suggest, moreover, that some *Al* electrons are transferred to *Mn*. We find that hybridization of orbitals is not predicted by analysis of the group theory of icosahedral symmetry. It therefore appears that some *Al* electrons are transferred to *Mn* by free electron metallic bonding and this contributes to the small volume, low enthalpy, and stabilization of the subcluster (appendix C).

Structure factors

Though the material contains many glassy defects (figure 7a), ideal structure factors can be calculated for 'logarithmically periodic' superclusters. Using the defined model of subcluster, cluster and superclusters, systematic structure factors for the various diffracted beams were computed. Bear in mind that in electron diffraction, the effect of separation of the reflecting sphere[13] from the plane of the reciprocal lattice is an additional factor that causes intensities to decrease with increasing scattering angle Θ. This decrease with increasing angle is further reduced by decreasing atomic scattering factors. For simplicity these two effects were treated as constant. With the advantage of centro-symmetry, quasicell scattering factors were then calculated using the standard formula[15] for crystals:

$$F_{hkl} = \sum_{j} f_j \cos(2\pi(hu_j + kv_j + lw_j))$$

(3)

but where a supercluster may be large, up to 250,000 atom sites for supercluster order 3. Here F_{hkl} is the structure factor for a quasi Bragg reflection with quasi Miller indices $h,k,l,$ while f_j is the scattering factor[13] for atom j having $u_j, v_j w_j,$ for its structural

78

coordinates in the model. Structure factors are usually calculated to identify forbidden lines in diffraction from crystals; we use them as an indication of diffracted intensity from quasicrystals. The modelled atoms were arranged in subclusters, clusters, and superclusters to orders 1, 2 and 3. The calculation was stopped at the last level for consistency with HREM imaging (figure 7). Part of this figure resembles the view of a compressed supercluster order 1 (figure 9a), showing atomic sites. Calculations were performed for all of the diffracted beams in the published C_{5v}, C_{3v} and D_{2h} patterns[1]. The calculations show:

1. The brightest ring in all of the C_{5v}, C_{3v} and D_{2h} patterns is on the circle C in figures 1-3.
2. The linear diagonal pattern in D_{2h} is strongly allowed; the Fibonacci series diagonal forbidden.
3. Confirming the Compromise Spacing effect, there is a systematic shift of about 5.6% in the position of diffraction peaks relative to interplanar spacings.
4. There is a conclusive match between calculated structure factors with rankings of experimental beam intensities in selected area diffraction (table IIIb).

The fact that, in the D_{2h} diagonals, Bragg diffraction occurs from planes that are separated linearly, and not from those separated by Fibonacci series spacings, suggests a new criterion for Bragg reflection in quasicrystals. Though the two types of planes are parallel and interleaved (figure 9a), the most obvious difference in the structures of the planes is this: whereas for the Fibonacci series patterns in C_{5v}, C_{3v} and D_{2h}, atomic planes are aligned in the directions of the plane normals; on diagonals in D_{2h} by contrast, the planes separated by Fibonacci series spacings, are misaligned. The dominance of the linear over Fibonacci series on the diagonals of the D_{2h} pattern, follows from a feature that describes the scattering of fast electrons. The feature can be represented by $\textbf{\textit{K.r}}_i$ = nd, summed over quasicells at locations $r_{i.}$ on a Bragg plane, where $\textbf{\textit{K}}$ is the scattering vector. In crystals, respective cells in Bragg planes are aligned along plane normals, as in those planes that cause Fibonacci series diffraction in C_{5v}, C_{3v} and the D_{2h} cross (figure 9). When, however, the *quasicells* are misaligned as on the D_{2h} diagonals, $\textbf{\textit{K.r}}_i$ adds destructive phases to the scattered waves. By contrast, on planes connected *within* quasicells (figure 9a), atoms are aligned parallel to the plane normals that point towards the direction of the diagonals in the pattern. The criterion therefore is: *In diffraction from quasicrystals, allowed reflections occur from Bragg planes that are aligned in the directions of the plane normals.* The various approaches are provided as an aid to understanding this complex D_{2h} pattern. Meanwhile, the forbidden line, listed as 'rotated E1' in table IIIb is significant. The line shows not only that forbidden lines are correctly represented in the calculations, but it also corroborates an apparent error (appendix D) in the original data of Schechtman et al.[1].

Corollary 3. Atomic planes that are misaligned in the direction of their normals do not diffract.

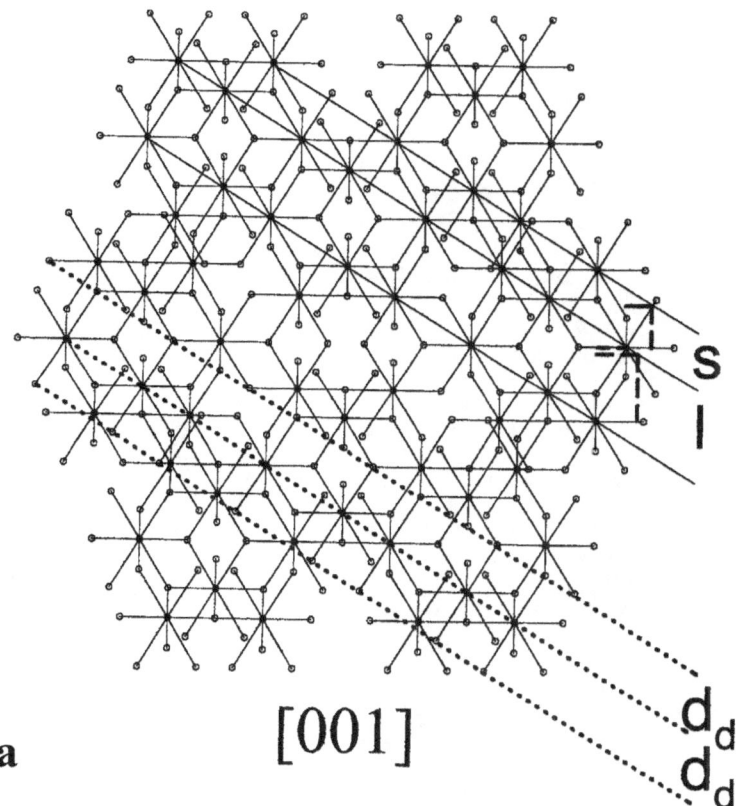

a   [001]

Figure 9. (a) View of supercluster order 1 from [001] axis. Filled circles are *Mn*; centred on open circles, *Al*. The intercellular Fibonacci series diagonal planes are misaligned; vertical and horizontal Fibonacci series planes, as also the intracellular linear diagonal planes (with spacing d_d) are aligned. The misaligned planes do not diffract.

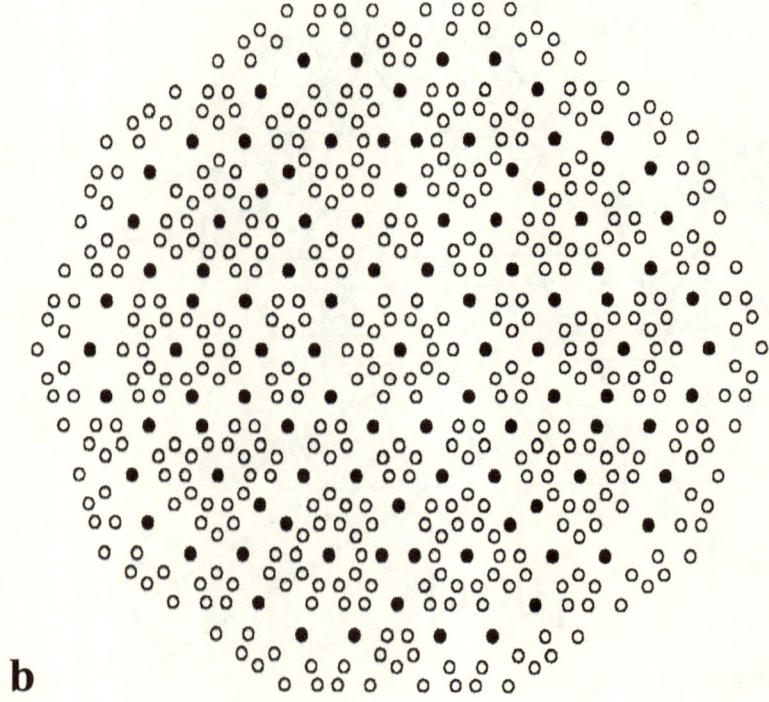

b

[01τ]

Figure 9 (b) 5-fold view from [01_τ], with aligned planes consistent with high-resolution images[8].

c

[111]

Figure 9 (c) 3-fold [111] view, with aligned planes.

Concluding this section on diffraction in quasicrystals, we point to preliminary speculations about their electronic band structures. By analogy with band structures in crystals, but applying logarithmic scales, a tentative solution for approximately free electrons is illustrated in appendix figure A.1 .

5.   Perspective and Summary

It is over 25 years since quasicrystals were discovered and over twenty years since fortuitous CBED was published.   In the intervening time many discoveries have been made and developments taken place. The developments described in this paper are not upstart. Our data was there near the beginning and we have returned to it privately and repeatedly since then. This is the first time they have been properly explained. Old data[5], that couldn't be explained at the time is as valid today as then. Subsequent models that had not had the benefit of the explanation are restricted in scope. New models that use old data are as valid now as they would have been then, if proposed earlier. There are features that are unique to our model the seeds of which have grown from the earlier paper[5].

- Firstly, there is the elucidation of the driving force. The icosahedral subcluster has a volume 17% less than the equivalent set of atoms in the matrix solution. The proof depended on the correct scaling for the icosahedral subcluster, including the physical effects that have been described.
- The physical effects include departures from Bragg's law of diffraction and an explanation of the geometric, Fibonacci series in spacings between diffracted beams.
- Prior to these developments, a convenient new method for indexing the pattern was derived by simple inspection. This

was facilitated by the observations, firstly of double diffraction at large angles only in the CBED, and secondly of the corresponding superposition of Bragg and Fibonacci components in the 2-fold diffraction pattern. Additionally, the selection by nature of the components that diffract had to be explained. The new indexation conveniently matched the scaling by representing the unit cell in reciprocal space with the unit cube in icosahedral units.

- The simple and complete indexation in turn facilitated the calculation of structure factors from superclusters. It had been noticed before that icosahedra can be aligned by edge sharing, but we developed this by calculating the structure factors for clusters of unlimited size and logarithmic periodicity.

- The model made from triadic golden rectangles facilitated the construction of coordinates for its constituent components.

- The calculations verified many features and included a discovery about interplanar spacings in quasicrystal diffraction. This reflected back onto the model in a critical way. Other details that had been overlooked, were observed.

- One of these was disparities from pure icosahedral symmetry in published data. The disparities are important for the information they give on expected defects the structure. Defects are thermodynamically enhanced in metastable alloys. General information about these defects can be obtained from optimum defocus HREM, and they can be modelled using our supercluster model as a skeleton basis.

- The model is consistent with optimum defocus HREM

This body of discoveries provides a consistent view of metastable quasicrystalline Al_6Mn. There are overlaps between our model and those of others, including some cited in the above work, but the conceptual framework is different: our understanding is unified by the single driving force, their alternative is complex and multitudinous.

The most important difference is their 'concept of quasicrystals', as a type of solid stabilized by vaguely conceived and novel atomic arrangements, supported by abstruse mathematics. Such views were cited in the opening five paragraphs of part II of this book. By contrast, our view is conventional. Other approaches, especially those engaged in puzzle building, have been neither one nor the other.

There is a particular feature in recent developments[9,16] that should make you wary. It is to do with orientational unrepeatability of dodecahedra while overlooking the chemical driver. The dodecahedron has the same point group symmetry as its reciprocal Platonic solid, the icosahedron. If you take the regular icosahedron, illustrated on the front cover of this book or in figure 4a, and place an atom over the centre of every triangular face, connecting their centres is a regular dodecahedron with pentagonal faces. The trouble is that, whereas the orientation of icosahedra are transmitted by edge sharing, such transmission cannot occur with dodecahedra. The model on page 31 shows that orientations of face shared dodecahedra vary unless considered perhaps in a single direction only, as if in a linear quasi molecule – not in a three dimensional quasicrystal. Edge sharing icosehedra therefore provide a simpler and more obvious explanation for the orientational symmetry observed in Al_6Mn quasicrystals – besides their demonstrated general consistency.

In summary, the model has unique explanatory and predictive power. (1) The driving force for the icosahedral structure is the high density of the icosahedral substructure. The supercluster model is (2) consistent with experimental data, where, in particular, high-resolution electron microscope (HREM) images show ring structures that fit dimensionally and geometrically, our clusters and first order supercluster. Moreover, we simulate (3) perfect icosahedral axial symmetries with a counter-intuitive bright ring C (typically, in electron diffraction, the brightest has smallest Θ), with (4) a description of the general features of Fibonacci series diffraction, including the absence of high order diffraction (excepting the diagonals in the D_{2h} pattern), and (5) double diffraction due partly to dynamical effects, confirmed experimentally by CBED. Meanwhile (6) the detailed features of the C_{5v} and C_{3v} Fibonacci series diffraction patterns are indexed, and (7) the D_{2h} axial pattern has been analysed and explained, using superposition between the two component Fibonacci and linear series. Conclusively, (8) the calculated structure factors match the ranked intensities of beams in recorded diffraction patterns.

The scope of this book focuses on the core icosahedral symmetry that was discovered in the first quasicrystals. It does not extend to related structures in materials science including the variety of compounds[17], the stability of phases[9], or directional asymmetries[18]. From the trunk will grow the branches.

There has been some hope that newly applied techniques, such as X-ray diffraction from comparatively large crystals would provide resolution of the quasicrystal structure   Because of their lower absorption rates, double diffraction is then less complicating than in electron diffraction, which can be applied to the smallest specimens.   Neutron diffraction requires yet larger specimens. However precedents are not good. It has long been known that X-ray diffraction provides ambiguous information on defective and amorphous structures, as in amorphous silica, owing to the phase problem.

There have been reports that the icosahedral second phase is dendritic[5]. This conclusion followed from an expectation due to segregation combined with rapid solidification. The conclusion was supported by circular particles observed in thin sections and an apparent parallelism between them. However, the consideration of growth mechanisms leads to a different expectation: spherical growth is suggested by the model. This is consistent with the imaging if there are preferred habit planes. The most obvious arise from similarities of axes when comparing cubic fcc with icosahedral symmetries, and more particularly with the separation of neighbouring *Al-Al* atoms. For example, given the prevalence of clusters in figure 7, the most likely fit is between the supercluster side length, 0.704 nm, with three times the fcc (111) interplanar spacing, 0.699 nm.

Predictably, the model will be used for understanding a range of physical, electronic and other properties, including also the wider fields of science that contain 5-fold symmetries. If the order of supercluster that was imaged in HREM is limited, remember that rapidly quenched materials favour defects. Moreover the image depends on the cuts of superclusters that occur within a foil with dominating decagonal and trigonal clusters. Meanwhile, the calculations prove the correctness of the solution to a long standing and fundamental structural problem.

Appendix

Appendix A. Schematic representation for the argument identifying the measured quasilattice parameter with half the golden ratio, $\tau/2$.

> Table A.I. Summary of derivation for model scaling factor from quasilattice parameter in electron diffraction patterns[*].

Rule	Explanation	Symbol definition	Derived values
$n\lambda = 2d\sin\theta$	Bragg's law for crystalline solids	n: order integer λ: wavelength d: interplanar spacing θ: Bragg angle	
$\approx d\Theta$	For high energy electron diffraction, $\lambda \ll d$	Θ: scattering angle	
$= \dfrac{a\Theta_{hkl}}{\sqrt{h^2 + k^2 + l^2}}$	For cubic unit cells	a: lattice parameter h,k,l: Miller indices in first order	
$\lambda\tau^m = d_m\,\Theta_m$	For Fibonacci spacing in diffraction from quasicrystals	m: integer = 1 τ: golden section or magic number = $\dfrac{1+\sqrt{5}}{2}$	
$d_m = a'/\tau^m$		a': quasilattice parameter	Make m=0 for the bright circle, c_0, with experimental d=0.206 nm = a' [*]
		s: scaling factor for icosahedral model of side $1s$ and rectangle length τs.	Suppose[†]: $\tau s/2 = 0.206$ nm So: $s = 0.255$ nm $\tau s = 0.412$ nm

* Before application of correction factor for Bragg-type reflection from quasicrystalline, Fibonacci series, interplanar spacings. The

first order correction is 1.0518, but the structure factor calculations include second order effects having the factor 1.056:

After correction: **subcluster side,** $s = 0.255 \times 1.056 =$ **0.269 nm**

Subcluster length, $l = 0.412 \times 1.056 =$ **0.435 nm**

The identification of lattice parameter, made in the table, is similar to that adopted by Tsai[9,10].

Appendix B. Summary for modified Bragg diffraction in quasicrystals.

It is commonly assumed that diffraction in quasicrystals, whether of electrons, x-rays or neutron*s,* is Bragg diffraction. Though there are some similarities (including scattering from atomic planes), there are also significant differences:

Bragg Diffraction was defined for crystals. Crystals are periodic under translation and contain orientational symmetries consistent with the fourteen Bravais lattices. Interplanar spacings are regular, e.g. $|d|d|d|\cdots$ Bragg diffraction is used to determine the structure of crystals. It results from constructive interference due to ordered planes of atoms. The diffraction follows Bragg's Law, $n\lambda = 2d\sin\theta$, for scattering of wavelength λ with order n and Bragg angle θ.

Quasicrystals display strict orientational symmetries without long range linear translational order. The symmetries are inconsistent with the Bravais lattices, and may contain 5–fold rotations. In the narrow sense, and in the short range, typical patterns can be explained by alternating periodicities of the type $|d|d/\tau|d|d/\tau|\cdots$, where the golden ratio $\tau = (1+\sqrt{5})/2$. Then $1/\tau + 1 = \tau$; $1 + \tau = \tau^2$ etc. Typically, but not always, quasicrystal diffraction patterns display spacings in Fibonacci sequences $1 : \tau : \tau^2 : \tau^3 : \ldots$

Such spacings are inconsistent with Bragg's law unless the **order** n is restricted to values of 0 or 1. The reason has been given that is consistent with constructive interference within the quasicrystals. This is satisfied only by scattered wave amplitudes between adjacent planes having phase relations of the type :

$$e^{2\pi i d/\lambda} . e^{2\pi i d/(\lambda\tau)} = e^{2\pi i d\tau/\lambda} .$$

(A.1)

Referring back to figure 8b, the diffraction condition in quasicrystals satisfies $\lambda = 2d\tau^{-m}\sin(\theta)$, where m is positive or negative integral. Filtering is provided by the relation $d\tau^{m-1} + d\tau^{m} = d\tau^{m+1}$ consistent with the Fibonacci series spacings.

Compare: in crystals, at small angles, $\theta \approx n\lambda/(2d)$; in quasicrystals, $\theta \approx \lambda/2d\tau^{-m}$. The exponent can be regarded as the logarithmic order

With the restriction in n, **double diffraction** along one dimension is neither simulated nor observed; but double diffraction is observed in the second dimension of the diffraction pattern. The last observation is abnormal, and is consistent with the last equation.

Consequently, in some diffraction patterns, as in the 2-fold pattern from Al_6Mn, **both Fibonacci sequences and linear sequences are evident and superposed.** Composite indexations, based on the unit cube in reciprocal space, allow remarkable agreement between calculated structure factors with observed diffraction beam intensities.

Which diffraction sequence is selected depends on the **alignment** of Bragg planes in the direction of the scattering vector. Misalignment results in incoherent scattering in the quasicrystals.

The **(Bourdillon) Compromise Spacing effect** , that is found both analytically and by simulation, provides a real quasilattice parameter that is larger than the corresponding Bragg interplanar spacing, d. This spatial effect is critical in fitting atoms into a theoretical structure.

As a consequence of the extraordinary diffraction in quasicrystals, their electronic **band-structures** are conveniently represented on graphs having logarithmic scales. For low-energy free-electrons, an example is shown in figure A1. Here h is Planck's constant, m the electronic mass, and k the wave-vector. The figure can be compared with band structures of crystals[19], in the extended zone scheme. Notice that whereas wave-functions in crystals superpose arithmetically; in quasicrystals, their phases add as in equation A.1.

Owing to the large number of sides in the icosahedral structure, the equivalent Brillouin zones and Fermi surfaces are closer to spherical than is typical in crystals. This provisional band structure is a basis for understanding electronic and other properties beyond the scope of this book.

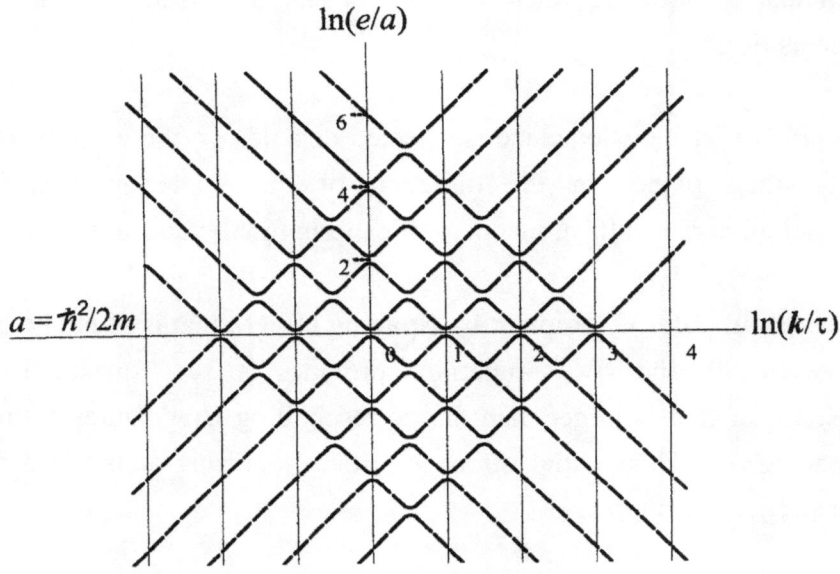

Figure A1. Provisional free-electron energy band-structure, drawn on logarithmic scales, in a quasicrystal.

Appendix C. Local stability in icosahedral subclusters.

Mn atoms have a diameter that is 12% smaller than in Al^{20}. Figure A2a illustrates its position between three close packed (111) planes in fcc *Al*. It is difficult for the surrounding atoms to relax, so that the Mn, shown by solid filling, has space to 'rattle.' The volume occupied by the atoms can be accurately calculated. The structure is sometimes represented by touching spheres. By contrast, the subcluster in figure A2b gives no space for rattle. The three-fold [111] axis lies normal to the planes shown. The view is similar to that in figure A2a, but with the further difference that the central plane is frilled: alternating atoms being above or below the central diameter through *Mn,* as indicated by arrows and by figure A3. Adjacent *Al* atoms are equispaced – both within planes and between them - being separated by the unit side length of the icosahedron. The relative sizes of *Al* and *Mn* atoms favour both the golden rectangle construction and the icosahedral subcluster. Thus it can be seen that, by including the *Mn* solute, the icosahedron, at this scale of solidification, is more compact than the fcc with *Mn* solute. This compaction is accompanied by a further measured reduction in scale. In consequence the volume of this subcluster is 17% smaller than the volume of a similar set of atoms in the fcc matrix. This reduction in volume in the quasicrystalline phase corresponds to

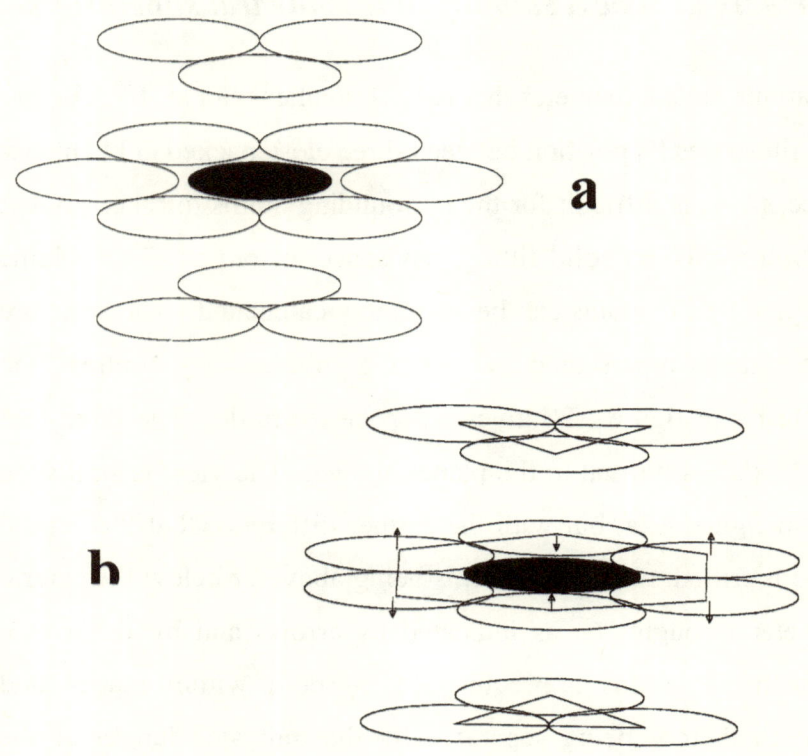

a

h

Figure A2. (a) Illustration of (111) planes of cross-sectioned atoms in fcc *Al*, containing dissolved *Mn*, in solid fill at the centre. Notice that the smaller size of the solute in the close packed structure gives room for 'rattle'.

(b) View of corresponding atoms in the icosahedral subcluster oriented with the three-fold (111) normal close to vertical. The atoms in the central plane through *Mn* are offset as indicated so that all adjacent Al atoms are separated by the same unit icosahedral side length. The central *Mn* atom is tightly bound with no rattle.

a reduction in enthalpy that stabilizes the subclusters and is the driving force for the phase.

A cluster scales as τ^2. The volume occupied by the cluster is, therefore, 17.94 times the volume of a corresponding subcluster, though the cluster is constituted from only 12 subclusters. The chief difference lies at their centres since the subcluster contains a tight central *Mn* atom; whereas the cluster centre is a combination *Al* atoms and vacancies. At this scale, the enthalpy of the supercluster appears to depend less on total volume; the enthalpy depends rather on the short *Al* bonds and on the defects and intergrowths described earlier. It thus appears that observed superclusters are the way that the energetically favourable subclusters are banded together in the metastable solid. Meanwhile the short *Al-Al* bond length suggests that metallic bonding and properties are enhanced within quasicrystal subclusters. This is consistent with group theoretical considerations that, we find, make hybridization of the atomic orbitals undefined within the icosahedral symmetry.

Appendix D. Orientation of 2-fold diffraction pattern.

Corollary 4. Schechtman's data[1] is not strictly icosahedral.

The original data of Schechtman et al. is not *in fact* icosahedral (figures A3 and A4), though the discrepancy could be a simple error in presentation. This has gone unnoticed for along time. It is necessary to know the true pattern, because divergences from pure symmetry cannot otherwise be investigated, and such divergences are especially important in directionally solidified materials. The effects of the fracture evident in figure 8a is an example of the asymmetry, since the fracture is normal to the direction of (quasi)crystal growth. Likewise, the model that was proposed in ref. 8 is asssymetric.

The following figure captions describe the anomaly, and the simulations show the expected diffraction patterns that are properly icosahedral. Schechtman et al. had represented the patterns at D_{2h}^{525} the same as at D_{2h}^{53235}.

Strange, isn't it – this was overlooked for 25 years? What else?

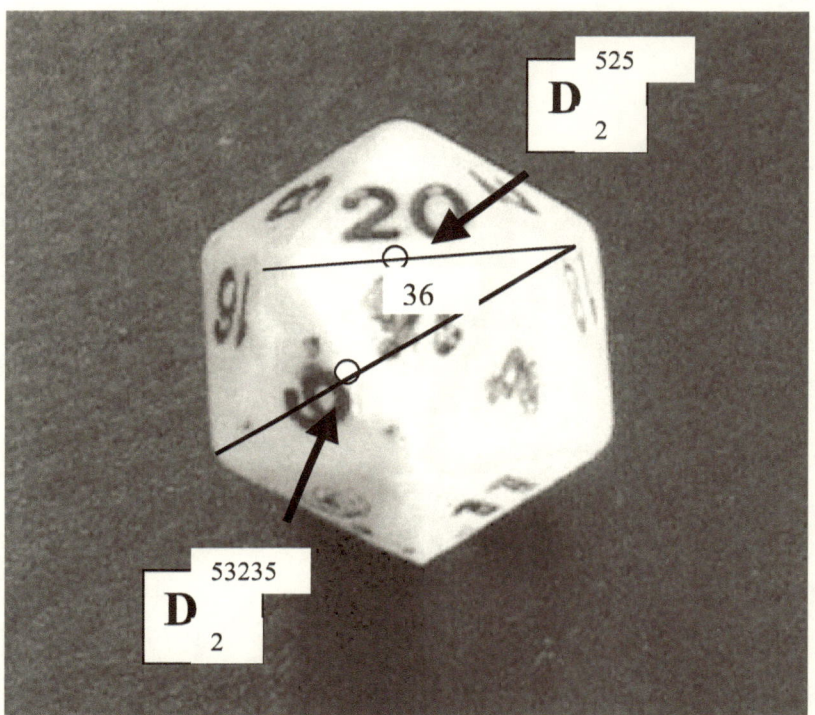

Figure A3. The pattern at D_2^{525} (the superscript identifies the axis of rotation on the stereographic projection) was reported[1] to rotate about D_{5v} by the angle 36^0 from D_{2h}^{53235} ; but should, by symmetry, rotate 54^0. In true icosahedra, the symmetry at D_{2h}^{525} is normal to the symmetry at D_{2h}^{53235}. The structure factor calculations (especially the bottom row of table IIIb) and model suggest an error in the published orientations[1] – D_{2h}^{525} should rotate by 90 degrees.

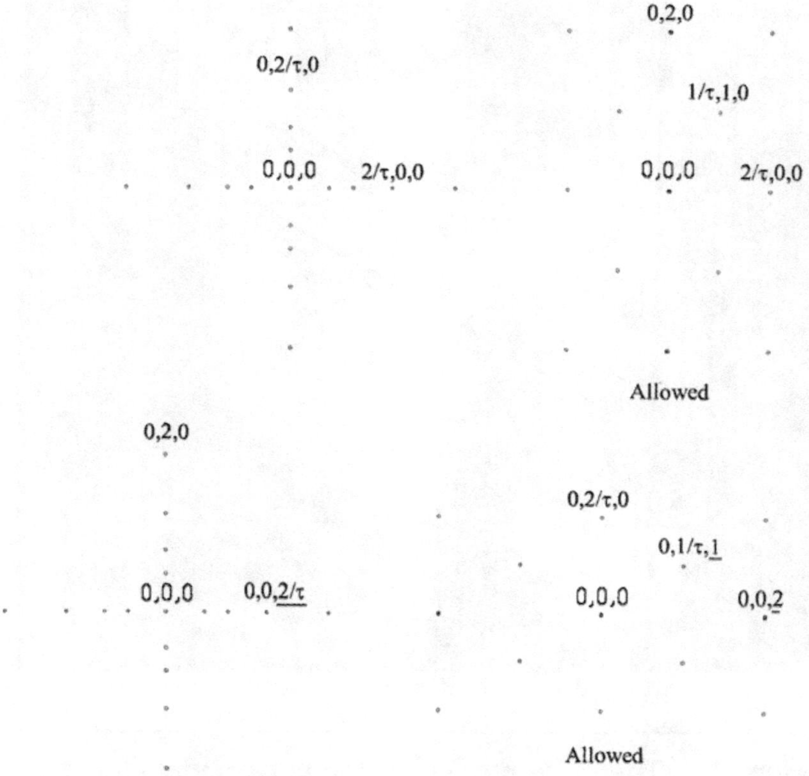

0,2,0

0,2/τ,0

1/τ,1,0

0,0,0   2/τ,0,0

0,0,0   2/τ,0,0

Allowed

0,2,0

0,2/τ,0

0,1/τ,1

0,0,0   0,0,2/τ

0,0,0   0,0,2

Allowed

a

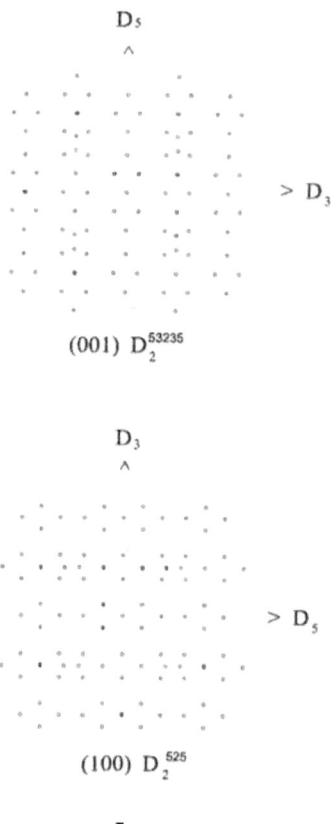

D₅
∧

> D₃

(001) D₂^{53235}

D₃
∧

> D₅

(100) D₂^{525}

b

Figure A4. (a) Comparison of allowed D₂ₕ diffraction patterns calculated for (001) and (100), consistent (b) with corresponding expected orientations (upper or lower) for true icosahedral symmetry, and different from D₂ₕ^{525} in ref. 1.

Index

Reference

[1] Schechtman, D., Blech, I., and Gratias, D and Cahn, J.W., *Metallic phase with long-range orientational order and no translational symmetry, Phys. Rev. Lett.*, **53**, p.1951 (1984)

[2] Senechal, M., *What is a Quasicrystal? Notices to the American Mathematical Society,* **53,** 886-887 (2006).

[3] Cahn, J.W., Schechtman, D, and Gratias, D., *Indexing of icosahedral quasiperiodic crystals, J. Mat. Res.* **1** 13-26 (1986)

[4] Elser, V., *Indexing problems in quasicrystal diffraction. Phys Rev. B*, **32** 4892-4898 (1985). See also ref. 9.

[5] Bourdillon, A.J., *Fine line structure in convergent-beam electron diffraction of icosahedral Al_6Mn, Phil. Mag. Lett.* **55**, 21-26 (1987)

[6] Huntley, H.E., *The Divine Proportion,* Dover, 1970.

[7] Pauling, L., *Apparent icosahedral symmetry is due to directed multiple twinning of cubic crystals, Letters to Nature,* **317,** 512-514 (1985). Unfortunately the author chose not to evaluate electron diffraction data.

[8] Bursill, L.A. and Peng, J.L., *Penrose tiling observed in quasi-crystal, Nature,* **316**, 50-51 (1985)

[9] Tsai, A.P., *Icosahedral clusters, icosahedral order and stability of quasicrystals—a view of metallurgy, Sci. Technol. Adv. Mat.* **9** 1-20 (2008)

[10] Takakura, H., Gomez, C.P., Yamamota, A., De Boissieu, M., and Tsai, A.P., *Nature Materials,* **6** 58-63 (2006).

[11] Levine, D. and Steinhardt, P.J., *Phys. Rev. B.* **34**, 596-616, (1986), see also Duneau, M. and Katz, A., *Phys. Rev. Lett.* **54,** 2688-2691 (1985).

[12] Buxton, B.F., Rackman, G.H., and Steeds, J.W., 1978, *Proceedings of the Ninth International Conference on Electron Microscopy,* Toronto, Vol. 1, pp188-189.

[13] Hirsch, P., Howie, A., Nicholson, R.B., Pashley, D.W., and Whelan, M.J., *Electron Microscopy of Thin crystals,* Krieger, 1977

[14] Reyes Gasga, J., Van Tendeloo, G. and Yacaman, M.J., in *Quasicrystals and incommensurate structures in condensed matter,* ed. Yacaman, M.J., Romeu, D, Castano, V. and Gomez, A., World Scientific (1989) pp 356-371.

[15] see e.g. Cullity, B.D., *Elements of X-ray diffraction,* Addison-Wesley, 1978.

[16] Forthcoming proceedings of ICQC9, Zurich, July 6-11, 2008.

[17] E. Macia, *The role of aperiodic order in science and technology, Rep. Prog. Phys.* 69(2006)397-441.

[18] A. Yamamoto, *Software package for structure analysis of quasicrystals, Sci. Technol. Adv. Mat.,* **9**, 1-14 (2008).

[19] Kittel, C., *Introduction to solid state physics*, Wiley, 1976.

[20] *CRC Handbook of Physics and Chemistry* (1990) ed. D.R.Lide, CRC, Boca Raton.

www.ingramcontent.com/pod-product-compliance
Lightning Source LLC
Chambersburg PA
CBHW022024170526
45157CB00003B/1346